Pomegranate

Edited by Vasiliki Lagouri

Published in London, United Kingdom

IntechOpen

Supporting open minds since 2005

Pomegranate
http://dx.doi.org/10.5772/intechopen.90501
Edited by Vasiliki Lagouri

Contributors
Sangram S. Sahebrao Dhumal, Ravindra D. Pawar, Sandip S. Patil, Amulya Thotambailu, Jiju Narayanan, Deepu Cheriamane, Satheesh Kumar Bhandary, Prakash Bhadravathi Ganesh, Manjula Santhepete, Sally Elnawasany, Vildan Celiksoy, Charles M. Heard, Vasiliki Lagouri, Maria Trapali

Notice
Statements and opinions expressed in the chapters are these of the individual contributors and not necessarily those of the editors or publisher. No responsibility is accepted for the accuracy of information contained in the published chapters. The publisher assumes no responsibility for any damage or injury to persons or property arising out of the use of any materials, instructions, methods or ideas contained in the book.

First published in London, United Kingdom, 2022 by IntechOpen
IntechOpen is the global imprint of INTECHOPEN LIMITED, registered in England and Wales, registration number: 11086078, 5 Princes Gate Court, London, SW7 2QJ, United Kingdom
Printed in Croatia

British Library Cataloguing-in-Publication Data
A catalogue record for this book is available from the British Library

Additional hard and PDF copies can be obtained from orders@intechopen.com

Pomegranate
Edited by Vasiliki Lagouri
p. cm.
Print ISBN 978-1-83968-447-0
Online ISBN 978-1-83968-464-7
eBook (PDF) ISBN 978-1-83968-465-4

We are IntechOpen,
the world's leading publisher of
Open Access books
Built by scientists, for scientists

6,000+
Open access books available

147,000+
International authors and editors

185M+
Downloads

Our authors are among the

156
Countries delivered to

Top 1%
most cited scientists

12.2%
Contributors from top 500 universities

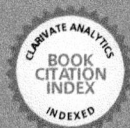

Interested in publishing with us?
Contact book.department@intechopen.com

Meet the editor

Vasiliki Lagouri, BA, MSc, Ph.D., received a BA in Chemistry from Aristotle University of Thessaloniki, Greece, where she was awarded by the National Fellowships Foundation, the highest undergraduate honor. She received an MSc in Medicinal Chemistry from the Department of Pharmacy, National and Kapodistrian University of Athens, and a Ph.D. in Food Chemistry from the Aristotle University of Thessaloniki. She was awarded post-graduate grants from the National Fellowships Foundation and the Mpodosakis Foundation of Greece. She has more than twenty years of research and academic experience at the Departments of Food Science and Technology, International Hellenic University of Thessaloniki, and the Departments of Chemistry and Pharmacy, National and Kapodistrian University of Athens. She has sixteen publications and twenty-four conference participations in the fields of food chemistry, medicinal chemistry, olive oil, and natural products to her credit.

Contents

Preface   XI

Section 1
Introduction   1

Chapter 1   3
Introductory Chapter: Pomegranate
by Vasiliki Lagouri

Section 2
Pomegranate Health Properties   9

Chapter 2   11
Antimicrobial Potential of Pomegranate Extracts
by Vildan Celiksoy and Charles M. Heard

Chapter 3   35
Role of Pomegranate in the Management of Cancer
by Amulya Thotambailu, Deepu Cheriamane, Manjula Santhepete,
Satheesh Kumar Bhandary, Jiju Avanippully and Prakash Bhadravathi

Chapter 4   47
Vasculoprotective and Neuroprotective Effects of Various Parts
of Pomegranate: In Vitro, In Vivo, and Preclinical Studies
by Maria Trapali and Vasiliki Lagouri

Chapter 5   59
Could Pomegranate Fight against SARS-CoV-2?
by Sally Elnawasany

Section 3
Post-Harvest Technology of Pomegranate   71

Chapter 6   73
Post-Harvest Management and Value Addition in Pomegranate
by Sangram S. Dhumal, Ravindra D. Pawar and Sandip S. Patil

Preface

Pomegranate (*Punica granatum L.*) is one of the oldest edible fruits in the Mediterranean and has been used extensively in folk medicine. The popularity of pomegranate has increased especially in the last decade because of its health effects. This book presents a comprehensive overview of pomegranate, its beneficial properties, and potential applications.

The introductory chapter, chapter 1, summarizes previous research on the anti-oxidant properties and bioactive polyphenolic ingredients in Greek pomegranate varieties. Chapter 2, "Antimicrobial Potential of Pomegranate Extracts", highlights the growing number of publications that have investigated the activity of pome-granate extracts against microbes. Research generally supports folklore claims and has shown that pomegranate extracts possess unusual and potent broad-spectrum activities against Gram-positive and Gram-negative bacteria (planktonic and biofilm), fungi, viruses, and parasites. Possible pathways/mechanisms of antimicrobial activity of pomegranate extracts are discussed and enhancement of such activity using metal ions is considered. The role of pomegranate in cancer treatment and as an adjuvant in therapy has been explored very little because there very few studies have been conducted on humans. However, there are a handful of animal and cell-line studies that deem the fruit and its extracts effective in the treatment of cancer. The studies conducted so far, as described in Chapter 3 "Role of Pomegranate in the Management of Cancer", show the potency of pomegranate and its components in the treatment of cancers of the prostate, breast, head and neck, colon, lungs, and skin or as an adjuvant in cancer treatment to minimize unwanted side effects. The various components of pomegranates, because of their antioxidant and anti-inflammatory properties, can be applied to various treatment strategies in numerous types of cancer in some way. Pomegranate is still surprising the world with its great therapeutic benefits. Chapter 4, "Vasculoprotective and Neuroprotective Effects of Various Parts of Pomegranate: In Vitro, In Vivo, and Preclinical Studies", in the section on "Pomegranate Health Properties", presents some recent studies on the vasculoprotective and neuroprotective effect of various parts of pomegranate and its main compounds, especially hydrolysable tannins ellagitannins, ellagic acid, and their metabolites. The in vitro and in vivo studies showed that the whole parts of pomegranate as well as its main components had a positive influence on blood glucose, lipid levels, oxidation stress, and neuro/inflammatory biomarkers. As such, they can be used as a future therapeutic agent for several vascular and neurodegenerative disorders such as hypertension, coronary heart disease, and Alzheimer's. Chapter 5 "Could Pomegranate Fight against SARS-CoV-2?", examines the potential properties of pomegranate against SARS-CoV-2 infection. Anti-viral and immune-modulating actions, besides tyrosine kinase and ACE inhibition potentials, all enable pomegranate to fight SARS-CoV-2 infection. In the last few years, the research and development activities on pomegranate fruit have aimed to develop technologies for new pomegranate-derived food products. As described in Chapter 6 "Post-Harvest Management and Value Addition in Pomegranate", in the section on "Post-Harvest Technology of Pomegranate", the pomegranate can be processed into products like minimally processed fresh arils, juice, squash, bever-ages, molasses, juice concentrates, frozen seeds, jam, jelly, marmalades, grenadine, wine, seeds in syrup, pomegranate spirits, pomegranate powder, pomegranate rind powder, anardana, confectionery, pomegranate seed oil, and more. These products are not yet popularized on a large scale due to a lack of commercially viable processing technologies. The modified atmosphere packaging offered an additional innovative tool for the optimal use and value addition of lower-grade pomegranate fruits. The minimally processed pomegranate arils and frozen arils packed in punnets and pomegranate juice are the most appealing products to consumers. Pomegranate juice can be used in beverages and jellies, for preparation of pomegranate juice concentrate, as a flavoring and coloring agents, and for dietetic and prophylactic treatment purposes. This new sector of pomegranate industrial processing will allow the use of non-commercial pomegranate fruits with some physical defects and fruit disorders to prepare new products, thus improving pomegranate utilization for human health.

Vasiliki Lagouri
Institute of Chemical Biology,
National Hellenic Research Foundation,
Athens, Greece

Section 1

Introduction

Chapter 1

Introductory Chapter: Pomegranate

Vasiliki Lagouri

1. Introduction

1.1 Antioxidant properties and bioactive polyphenolic ingredients in pomegranate varieties

Pomegranate (*Punica granatum* L.) has been used extensively in folk medicine of many cultures. The popularity of pomegranate has increased tremendously especially in the last decade due to its antimicrobial, anti-viral, anti-cancer, powerful antioxidant and anti-mutagenic effects of the fruit [1–4]. Pomegranate juice has been proposed as a chemopreventive, chemotherapeutic, antiathero-sclerotic and anti-inflammatory agent. Polyphenols, which represent the predominant class of phytochemicals in pomegranate, consist mainly of hydrolyzable tannins such as gallotanins, ellagitannins and ellagic acid, components which have high antioxidant activities [5–8].

These activities are attributed to the pomegranate's high levels of polyphenols content. Polyphenols, represent the predominant class of phytochemicals of pomegranate fruits, mainly consisting of hydrolysable tannins which are mainly located in the fruit peel and mesocarp of pomegranates. Chemical analyses have shown that the pomegranate (juice) contains a significantly high level of hydrolysable tannins, such as gallotannins, ellagitannins and ellagic acid, which exhibited high antioxidant activities [9].

Ellagic acid has been found to exhibit antimutagenic, antiviral, whitening of the skin and antioxidant properties. EA exhibits significant anti-mutagenic, antitoxic, anti-apoptotic, anti-cancer, antibacterial, antiviral, anti-diabetic and anti-inflammatory properties [10–12].

Urolithins are products of metabolism of ellagic acid through the loss of one of the two lactones (lactonase/decarboxylase action) and by successive hydroxyl removal (dehydroxylase activities). In vitro trials have shown anti-inflammatory, anti-cancer, anti-glycemic, antioxidant and antimicrobial effects of urolithins, supporting their potential health benefits attributed to foods rich in pomegranate and ellagitannins [13].

The antioxidant properties and the presence of antioxidant compounds has been reported mostly for pomegranate juice [14–16] however, increasing literature was found reporting the antioxidant activity of pomegranate peels and seeds [16–18].

The extensive knowledge about pomegranate's health attributes and public awareness about nutritional food has increased the demand for the industrial use of pomegranate fruit and its byproducts (peels and seeds). The peels and seeds, which are usually disposed of as waste material in many food-processing industries, could be a rich source of beneficial phytochemicals [19].

Methanol and/or combinations of methanol and other organic solvents have been used for the extraction of polyphenols from pomegranate peels [17, 18]. It is also important to study the use of water as an alternative solvent for the extraction of polyphenols because it is easily accessible, non-toxic, environmentally friendly, and non-hazardous to operator health.

2. Previous studies: Results

A study presented by Lagouri et al., in the 6th International Conference on "Oxidative Stress-Skin Biology and Medicine" 2014 with the title: "Antioxidant properties and phenolic content of Greek pomegranate cultivars" was performed in order to quantify total phenols, flavonoids, hydrolyzable tannins and ellagic acid in the juice, and in the peel and seed extracts of two pomegranate varieties from mainland Greece (Central Macedonia and Thrace) with high pressure liquid chromatography method.

The aims of the study were to prepare aril juices, peel homogenates, peel and seed aqueous and methanolic extracts of two pomegranate cultivars collected from mainland Greece (Central Macedonia: B cultivar and Thrace: C cultivar) and to evaluate:

- their antioxidant properties by using free radical scavenging (DPPH) and ferric reducing antioxidant power (FRAP) assays

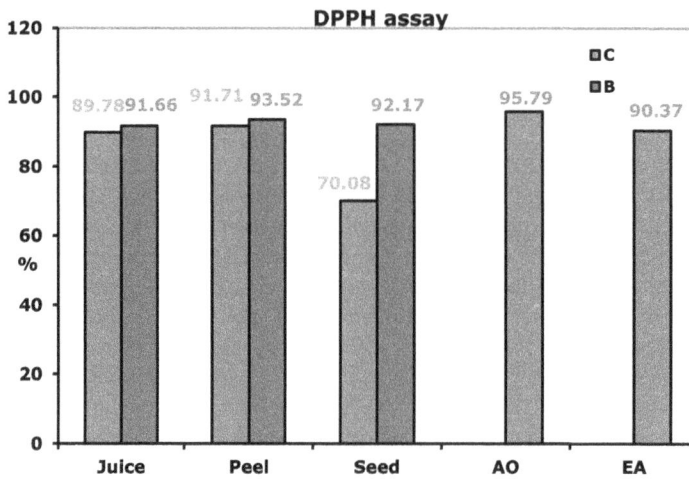

Figure 1.
DPPH assay.

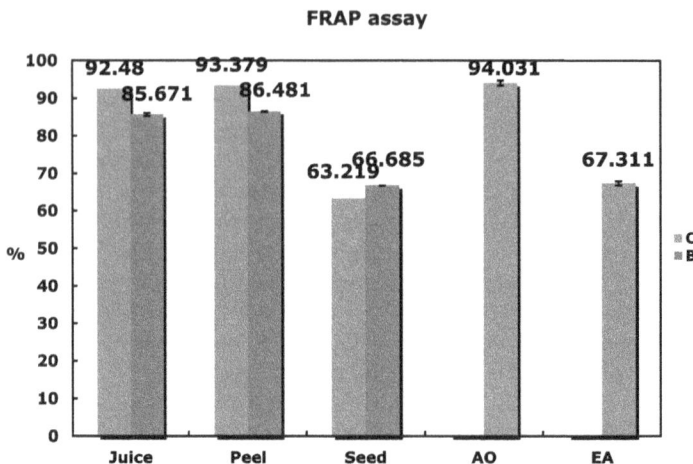

Figure 2.
FRAP assay.

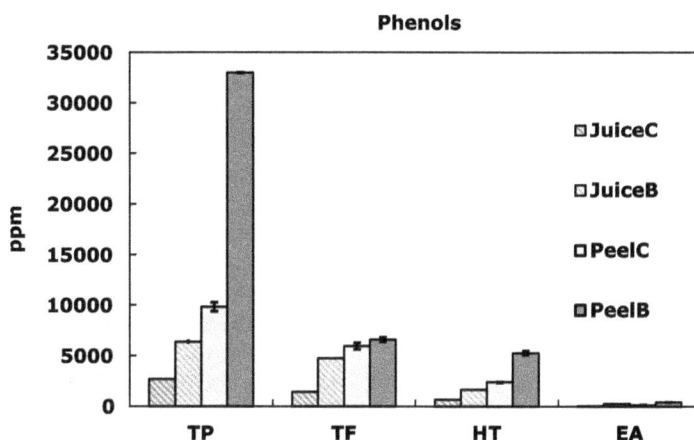

Figure 3.
Total phenols assay.

- their total phenols (TP), total flavonoids (TF), hydrolysable tannins (HT) and ellagic acid (EA) contents by using spectrophotometric and high pressure liquid chromatographic methods [20–24].

In total as shown in **Figures 1–3**, the peels from both pomegranate cultivars had higher antioxidant activity and phenol contents compared to juices and seeds. In addition, the B cultivar (Central Macedonia) in its peel homogenates showed higher free radical scavenging activity, total phenol, total flavonoid, hydrolysable tannins and ellagic acid contents than the C cultivar (Thrace). From the results it can be concluded also that different solvents (methanol, water) at temperatures 24 and 40°C used during the extraction process of the peels may affect their antioxidant properties and phenol contents.

In conclusion the results of the study were very promising because pomegranate peels and seeds, which are commonly disposed of as waste in many food processing industries, could be important sources of phytochemicals.

Author details

Vasiliki Lagouri
Institute of Chemical Biology, National Hellenic Research Foundation,
Athens, Greece

*Address all correspondence to: vlagouri@eie.gr

IntechOpen

References

[1] Lansky EP, Newman RA. *Punica granatum* (pomegranate) and its potential for prevention and treatment of inflammation and cancer. Journal of Ethnopharmacology. 2007;**109**:177-206

[2] Vučić V, Grabež M, Trchounian A, Arsić A. Composition and potential health benefits of pomegranate: A review. Current Pharmaceutical Design. 2019;**25**(16):1817-1827. DOI: 10.2174/138 1612825666190708183941

[3] Wang D, Özen C, Abu-Reidah IM, Chigurupati S, Patra JK, Horbanczuk JO, et al. Vasculoprotective effects of pomegranate (*Punica granatum* L.). Frontiers in Pharmacology. 2018;**9**:544. DOI: 10.3389/fphar.2018.00544

[4] Zarfeshany A, Asgary S, Javanmard SH. Potent health effects of pomegranate. Advanced Biomedical Research. 2014;**3**:100. DOI: 10.4103/2277-9175.129371

[5] Çam M, Yasar H, Gökhan D. Classification of eight pomegranate juices based on antioxidant capacity measured by four methods. Food Chemistry. 2009;**112**:721-726

[6] Danesi F, Ferguson LR. Could pomegranate juice help in the control of inflammatory diseases? Nutrients. 2017;**9**(9):958. DOI: 10.3390/nu9090958

[7] Halliwell B, Gutteridge JMC, Cross CE. Free radicals, antioxidants and human disease: Where are now? The Journal of Laboratory and Clinical Medicine. 1992;**119**:598-619

[8] Scalbert A, Manach C, Morand C, Remesy C. Dietary of polyphenols and the prevention of diseases. Critical Reviews in Food Science and Nutrition. 2005;**45**:287-3065

[9] Seeram NP, Adamsa LS, Henninga SM, Niu Y, Zhang Y, Nair MG, et al. In vitro antiproliferative, apoptotic and antioxidant activities of punicalagin, ellagic acid and a total pomegranate tannin pomegranate juice. Journal of Nutritional Biochemistry. 2005;**16**(6):360-367

[10] Derosa G, Maffioli P, Sahebkar A. Ellagic acid and its role in chronic diseases. Advances in Experimental Medicine and Biology. 2016;**928**: 473-479

[11] Marella S, Hema K, Shameer S. Nano-ellagic acid: Inhibitory actions on aldose reductase and α-glucosidase in secondary complications of diabetes, strengthened by in silico docking studies. Biotech. 2020;**10**(10):439

[12] Vattem DA, Shetty K. Biological functionality of ellagic acid: A review. Journal of Food Biochemistry. 2005;**29**(3):234-266

[13] Espín JC, Larrosa M, García-Conesa MT, Tomás-Barberán F. Biological significance of urolithins, the gut microbial ellagic acid-derived metabolites: The evidence so far. Evidence-based Complementary and Alternative Medicine. 2013;**2013**:270418

[14] Gil MI, Tomas BFA, Hess PB, Holcroft DM, Kader AA. Antioxidant activity of pomegranate juice and its relationship with phenolic composition and processing. Journal of Agricultural and Food Chemistry. 2000;**48**:4581-4589

[15] Tezcan F, Gültekin ÖM, Diken T, Özçelik B, Bedia EF. Antioxidant activity and total phenolic, organic acid and sugar content in commercial pomegranate juices. Food Chemistry. 2009;**115**:873-877

[16] Tzulker R, Glazer I, Bar II, Holland D, Aviram M, Amir R. Antioxidant activity, polyphenol content, and related compounds in

different fruit juices and homogenates prepared from 29 different pomegranate accessions. Journal of Agricultural and Food Chemistry. 2007;**55**:9559-9570

[17] Singh RP, Chidambara MKN, Jayaprakasha GK. Studies on the antioxidant activity of pomegranate (*Punica granatum*) Peel and seed extracts using in vitro models. Journal of Agricultural and Food Chemistry. 2002;**50**:81-86

[18] Li Y, Guo C, Yang J, Wei J, Xu J, Cheng S. Evaluation of antioxidant properties of pomegranate peel extract in comparison with pomegranate pulp extract. Food Chemistry. 2006;**96**: 254-260

[19] Negi PS, Jayaprakasha GK, Jena BS. Antioxidant and antimutagenic activities of pomegranate peel extracts. Food Chemistry. 2003;**80**:393-397

[20] Benzie IF, Strain JJ. The ferric reducing ability of plasma as a measure of "antioxidant power" the FRAP assay. Analytical Biochemistry. 1996;**239**:70-76

[21] Brand-Williams W, Cuvelier ME, Berset C. Use of free radical method to evaluate antioxidant activity. Lebensmittel-Wissenschaft und -Technologie. 1995;**28**:25-30

[22] Slinkard K, Singleton VL. Total phenol analysis: Automation and comparison with manual methods. American Journal of Enology and Viticulture. 1997;**28**:49-55

[23] Willis RB. Improved method for measurement hydrolysable tannins using potassium iodate. Analyst. 1998;**123**:435-439

[24] Zhisen J. The determination of flavonoid contents in mulberry and their scavenging effects on superoxide radicals. Food Chemistry. 1999;**64**: 555-559

Section 2

Pomegranate Health Properties

Chapter 2

Antimicrobial Potential of Pomegranate Extracts

Vildan Celiksoy and Charles M. Heard

Abstract

The search for plant extracts with efficacious antimicrobial activity remains important, partly due to fears of the side effects associated with conventional antibiotics and to counter the emergence of resistant microorganisms. Pomegranate extracts have been used for millennia for their anti-infective properties, with activity more recently being attributed to its rich composition of ellagitannins and other secondary polyphenolic compounds. This chapter highlights the growing number of publications that have probed the activity of pomegranate extracts against microbes. Research generally supports folklore claims and has shown that pomegranate extracts possess unusual and potent broad-spectrum activities against Gram-positive and Gram-negative bacteria (planktonic and biofilm), fungi, viruses and parasites. Possible pathways/mechanisms of antimicrobial activity of pomegranate extracts are discussed and enhancement/potentiation of such activity using metal ions considered.

Keywords: antimicrobial, bacteria, fungi, viruses, parasites, polyphenols, pomegranate extracts, biofilm, tannins, punicalagin

1. Introduction

Infectious diseases caused by pathogenic microbes are a fundamental problem and remain one of the major factors behind high morbidity and mortality across the world, especially in developing countries. This is exacerbated by the world-wide emergence of antibiotic-resistant pathogens which has in turn given increased urgency to the discovery of new antimicrobial compounds, including those derived from plants [1, 2].

The pomegranate, fruit of the *Punica granatum* L. tree, is one of oldest recorded edible fruits and it has been used as a folklore medicine since ancient times. There are records of it being used to treat inflammatory diseases and disorders of the digestive tract in the Ayurvedic and Unani systems [3, 4]. In terms of infections, the ancient Egyptians used it in the treatment of tapeworms and other parasites [5], whereas other cultures have used pomegranates to treat diarrhea and dysentery [6–8], although at the time they would not have known that pathogenic microbes were responsible. In more recent times, the pomegranate has been extensively and scientifically studied for its antimicrobial potential in a diversity of areas such skin infections, dentistry, food preservation etc. [9].

The phytochemistry of pomegranate extracts is well described in the literature [10–12] and they are known to be rich in bioactive compounds especially polyphenolics including anthocyanins and ellagitannins, in particular punicalagin, which is in the highest proportion [13]. As will be seen, it has become apparent that the pomegranate possesses unusual broad-spectrum potency against a wide range of species, which generally correlates with its polyphenol concentration.

In this chapter we aim to summarise published research into pomegranate extracts as antimicrobials and discuss some of the purported mechanisms behind such activity. Finally, the enhancement of antimicrobial activity by co-administration with metal ions is considered.

2. Activity against bacteria

Staphyllococcus aureus (S. aureus) and methicillin resistant *Staphyllococcus aureus* (MRSA) have received the greatest attention as targets for pomegranate extract activity. In 2010, the antibacterial activity of crude and purified extracts of pomegranate peel were assessed by Panichayupakaranant *et al.* 8 mg crude peel loaded discs showed 20 mm and 30 mm zone of inhibition against clinical isolates of *S. aureus* and *E. coli*, respectively. The purified peel extract discs, loaded up to 8 mg, exerted a range of zones of inhibition between 15-20 mm for *S. aureus* and 20-30 mm for *E. coli*. Using standardized peel extract, minimum inhibitory concentrations (MIC) values of 0.016, 0.008, and 0.008–0.016 mg/mL were obtained for *S. aureus*, *S. epidermidis* and *Propionibacterium acnes* respectively. Tetracycline was used as a positive control in this study and standardized pomegranate rind extract showed lower activity in zone of inhibition assays, with tetracycline also showing a lower minimum inhibitory concentration (MIC) [14]. A methanolic extract of pomegranate peel inhibited biofilm formation and eradicated pre-formed biofilm of *S. aureus*, MRSA, *E. coli* in the concentration range 25 to 150 µg/mL [15]. In the same study, ellagic acid showed biofilm inhibition and eradication activity at somewhat lower concentrations (5–40 µg/mL) than pomegranate peel extract. Furthermore, while pomegranate extract was able to inhibit the growth of *S. aureus*, it also suppressed enterotoxin production [5].

Pomegranate extracts have shown antimicrobial activity against to a range of oral microbes. It has been found that pomegranate extract powder at 1 mg/mL was effective against primary and secondary colonizer bacteria of dental plaque: *F. nucleatum, P. gingivalis, P. intermedia, S. mutans* and *A. actinomycetomomitans* [16]. In another *in vitro* study, pomegranate alcoholic extracts have been tested on bacteria which are collected from patients who have tooth decay or periodontitis and inhibited a range of bacteria in both planktonic and biofilm conditions [17]. Synergistic bactericidal activity against *S. mutans* and *R. dentocariosa* was reported for pomegranate extract in combination with other plant polyphenolic extracts, honey and myrtle [18].

Moreover, 'standardized' pomegranate peel extract showed higher antimicrobial activity than other parts of pomegranate (flower, leaf, stem) and ciprofloxacin (2 mg/mL) against *S. mutans, Salmonella mitis* and *L. acidophilusin* in a zone of inhibition assay [19]. Again, pomegranate gel showed an inhibitory activity against *S. mutans, Salmonella sanguis,* and *S. mitis* [20]. This gel also showed antiadhesive activity against *S. mutans* and *S. mitis* at lower than minimum inhibitory concentrations to a glass surface. In addition to inhibition activity on bacterial growth and biofilm, pomegranate extracts showed antiadhesive activity for *S. mutans* adherence on tooth surface in orthodontic treated patients [21]. In other clinical studies, the antiplaque effect and prophylactic benefits of pomegranate have been highlighted [22]. Recently, a systematic review and meta-analysis has been carried out by Martins *et al.* [23], where natural antimicrobial phenolic compounds were

Part of pomegranate used	Form	Test organisms	MICs	Reference
	PomElla® (30% punicalagin)	*Streptococcus mutans,* *Fusobacterium nucleatum,* *Aggregatibacter actinomycetomcomitans,* *Prevotella intermedia*	0.8 μg/mL 0.2 μg/mL 0.2 μg/mL 0.8 μg/mL	[16]
Peel	Acetonic, Methanolic, Ethanolic PPE and Hydro-alcoholic PPE	*Streptococcus mutans,* *Gemella morbillorum,* *Enterococcus faecalis,* *Staphylococcus epidermis,* *Klebsiella oxytoca,* *Enterobacter bugandensis,* *R. dentocariosa,* *Streptococcus mutans*	0.0125–0.025 mg/mL 12.5–25 mg/mL 0.0125–0.025 mg/mL 0.05–0.4 mg/mL 12.5–25 mg/mL 3.15–100 mg/mL 10 mg/mL 10 mg/mL	[17, 18]
Peel, flower, leaf, stem	Aqueous and methanolic	*S. mutans, S. mitis,* *L. acidhopillus*		[19]
Peel	Basic gel formulation including 540 mg pomegranate peel powder	*S. mutans,* *S. mitis,* *S. sanguis,* *C. albicans*	1:16 1:128 1:16 1:64	[20]
	Pomegranate mouthwash (Pomegranate extract were obtained from Verdure science, 30% punicalagin)	*Aggregatibacter actinomycetemcomitans,* *Porphyromonas gingivalis,* *Prevotella intermedia*	62.5 mg/mL >31.25 mg/mL 16.125 mg/mL	[21]
Peel	Methanolic PPE	*S. aureus,* *MRSA,* *E. coli,* *C. albicans*	250 μg/mL 250 μg/mL 250 μg/mL 1000 μg/mL	[15]
Fruit pericarp	Methanolic	*Staphylococcus aureus,* *Streptococcus pyogenes,* *Escherichia coli,* *Klebsiella pneumoniae,* *Proteus vulgaris,* *Pseudomonas aeruginosa*	640–2560 μg/mL 1280–2560 μg/mL 640–2560 μg/mL	[9]
Peel	Standardized extract (13% ellagic acid)	*Propionibacterium acnes,* *Shigella sonnei,* *S. aureus,* *Staphylococcus epidermidis*	15.6 μg/mL 7.81 μg/mL 7.81–15.6 μg/mL 7.81 μg/mL	[14]
Peel	Water	*B. cereus, S. aureus,* *P. fluorescens, S. tyriphium,* *E. coli*	200–450 ppm	[24]
Peel	Acetonic, methanolic, water	*Bacillus cereus,* *Bacillus coagulens,* *B. subtilis,* *Staphylococcus aureus,* *E. coli,* *Pseudomonas aeruginosa*	200–400 ppm 150–500 ppm 200–450 ppm 200–700 ppm 200–400 ppm	[25]
Fruit	Water	*Bacillus cereus,* *E. coli,* *S. aureus*	100 mg/mL 100 mg/mL 100 mg/mL	[27]

Part of pomegranate used	Form	Test organisms	MICs	Reference
Peel	Ethanolic, methanolic	*L. monocytogenes*		[28]

Table 1.
Antibacterial activity of pomegranate extracts against different bacteria.

compared with synthetic antimicrobials by using 16 clinical studies for qualitative analysis, and 12 studies for meta-analysis. For the meta-analysis, six clinical trials were evaluated for the comparison of natural antimicrobial phenolic compounds, including pomegranate extract mouthwash, and synthetic antimicrobials. It was found that natural antimicrobial phenolic compounds are less effective than chlorhexidine for biofilm control, although it showed similar reduction of the oral microbes count which was sub-grouped as total microorganisms, *Streptococcus mutans*, and *Streptococcus spp.* according to type of microorganisms.

Due to its antimicrobial and antioxidant properties, pomegranate extract has been studied for its preservation potential use in the food industry. Kannat *et al.* [24] did a study to evaluate the antimicrobial activity of pomegranate peel against common food spoilers and potential pathogens. It was shown that pomegranate peel extract increased the shelf-life of chicken and meat products and showed antimicrobial activity against to *S. aureus, B. cereus* with a minimum inhibitory concentration at 0.01%. However, it did not show antimicrobial activity for *E. coli* and *S. typhimurium* even at higher concentrations. Other researchers showed that pomegranate extracts were less effective against Gram-negative compared to Gram-positive bacteria, probably due to the differences in cell wall structure [25, 26]. Moreover, pomegranate has been studied in a novel and smart multi-functional hydrogel (MFH) system as a food packaging material since it is easy to monitor of the color change due to changes in conditions such as pH and temperature. The MFH with pomegranate extract showed promising antimicrobial activity on pasteurized milk and cheese over a 7-day period [27]. Pomegranate peel extract was also added in a film formulation to produce a material for food packaging materials with antimicrobial and antioxidant effects. This film formulation was found to restrict the growth of *L. monocytogenes* in pork samples inoculated with this bacterium [28].

The activity of pomegranate extract against bacteria is summarized in **Table 1**.

3. Activity against fungi

Treatment of fungal infections is a big challenge because of the eukaryotic nature of fungal cells that have similarity with host cells. While there are some drugs in the treatment of fungal infections available in the clinic, they are limited and there is a need for new alternatives [29, 30]. There are reports showing antifungal activity of pomegranate extracts, especially against *Candida* species [5, 7, 31], which are part of the normal microbiota of human gastrointestinal, oral, and vaginal mucosae. However, they can cause superficial infections and especially in immunosuppressed patients, they can cause severe infectious problems. In one study, punicalagin showed superior antifungal activity than the conventional fluconazole in an *in vitro* time-kill assay. In addition, punicalagin caused a significant change in *Candida* morphology, and alteration in budding pattern and pseudo hyphae when yeasts were treated with a sub-inhibitory concentration of punicalagin [32]. In another study, pomegranate extract showed superior inhibitory action against *C. tropicalis* while fluconazole and voriconazole, which are commonly prescribed azoles for fungal infections, have been ineffective against *C. tropicalis* [33].

The potential of pomegranate extract has been studied against fungi in *in vitro* biofilm assays. Microbes in biofilms have substantially different characteristics to those of their free-living planktonic counterparts [34, 35]. In particular, microbes in biofilms are concealed and therefore protected from antifungal agents and plant extracts [36, 37]. Pomegranate extract and its one of the major components ellagic acid were shown to exert a reduction in biofilm formation and eradicated pre-formed biofilm of *C. albicans* in an *in vitro* biofilm study [15]. Spray-dried microparticles containing pomegranate extract showed antifungal activity in *in vitro* assays under both planktonic and biofilm conditions [38]. In addition, the inhibitory effect of pomegranate extract on the growth of *Candida albicans* was demonstrated in an *in vivo* study [31]. In another study, similar effects have been obtained against to *Candida mycoderma* using different parts of pomegranate, fresh fruit and sterile juice [39].

In addition to *Candida* species, pomegranate extracts showed inhibitory activity against dermatophytes, which are fungi which use keratin as a source of nutrition and may cause infection in keratinized tissue parts such as nails, skin and hair follicles. Pomegranate peel extract and punicalagin exerted potent antifungal activity against to *T. rubrum* (125 µg/mL), *T. mentagrophytes* (125 µg/mL), *M. canis* (250 µg/mL) and *M. gypseum* (250 µg/mL). Punicalagin at a concentration of 62.5 µg/mL also inhibited *T. rubrum* spore germination, and it was further found that punicalagin 62.5 µg/mL and nystatin 0.78 µg/mL showed similar inhibition in hyphal growth of *T. rubrum* [40]. Moreover, pomegranate extracts have been researched for use as natural preservatives due to their antifungal (in addition to antibacterial) activities [41]. Pomegranate peel extract showed inhibition activity against to *Aspergillus*

Part of pomegranate used	Form	Test organisms	MICs	Reference
Peel	Crude extract, Aqueous fraction, ethyl acetate fraction, butanol fraction, punicalagin	*C. albicans, C. parapsilosis*	3.9–7.8 µg/mL, 1.9–15.6 µg/mL	[32]
Peel	Methanolic	*C. albicans*	> 1000 µg/mL	[15]
Peel	Crude extract, Crude extract in a spray-dried microparticle formulation	*C. albicans, C. parapsilosis, C. tropicalis*	3.9–15.6 µg/mL	[38]
Peel	Hydroalcoholic	*Trichophyton rubrum, Trichophyton mentagrophytes, Microsporum gypseum, Microsporum canis*	125 g/mL, 125 g/mL, 250 g/mL, 250 g/mL	[40]
Peel	Aqueous	*Aspergillus niger, Aspergillus parasiticus*		[42]
Peel	Aqueous	*A. alternata, S. botryosum, Fusarium spp.*		[43]

Table 2.
Antifungal activity of pomegranate extracts against different fungi.

niger and *Aspergillus parasiticus* in a zone of inhibition assays [42]. Pomegranate extract showed inhibitory effects against fungal pathogens which are responsible for fruit and vegetable decay. Punicalagin was proposed as the main compound in the extracts providing the observed antifungal activity and it has been found effective in mycelial growth inhibition against phytopathogenic filamentous fungi such as *Fusarium vertillicoides, Mucor indicus, Penicillium citrinum, Rhizopus oryzae* and *Trichoderma recei* [43]. Also, the growth rate of pathogens presented a negative correlation with total punicalagin content, and it has thus been suggested that pure punicalagin may be used as a control agent in storage disease to prevent the excessive use of synthetic fungicides [44, 45].

The activity of pomegranate extract against fungal microbes is summarized in **Table 2**.

4. Activity against viruses

Pomegranate extracts have been examined as an alternative treatment for viral infections [46–48]. A number of studies have shown that polyphenolic compounds have broad-spectrum antiviral activity, by inhibiting viral DNA and RNA, and directly binding the viral particles. It has also been suggested that polyphenols could provide antiviral activity during intracellular replication [49–52].

Pomegranate peel extract showed antiviral activity against the influenza virus. In a study by Sundararajan *et al.* [53], complete inactivation of influenza virus was observed with 1600 µg/mL pomegranate polyphenols, and 400 µg/mL of same extract showed 99% or more titer reduction in only 5 minutes treatment. This result was similar to another study which showed complete inactivation of H3N2 influenza virus within 30 minutes of treatment and a significant viral reduction with approximately 1 µg/mL pomegranate polyphenols. An *in vitro* study, showed that pomegranate polyphenol extract inhibited viral replication in addition to its virucidal effect – they also obtained same activity for punicalagin and suggested punicalagin is the main compound in pomegranate extract for antiviral activity [54]. In an *in vivo* mouse model study, pomegranate polyphenols applied to the lung were found to reduce influenza infection, without toxic effect to the host [55, 56].

Hepatitis C virus (HCV) is the main factor in end-stage liver disease and approximately 170 million people are chronically infected with HCV. Pomegranate ellagitannins, punicalagin, punicalin and ellagic acid, blocked and inhibited the NS3/4A protease which is a viral polyprotein responsible for processing and replication in HCV. Moreover, punicalagin and punicalin significantly decreased the HCV replication in an *in vitro* cell culture system [57]. The more prevalent adenovirus (ADV) in Hep-2 host cells has also shown susceptibility to pomegranate crude extract, fractions, and main phenolic compounds. It has been found that a n-butanol fraction of pomegranate peel extract and gallic acid showed the highest antiviral activity against ADV. Furthermore, the crude extract, n-butanol fraction and gallic acid inhibited ADV replication in the post-adsorption phase [58].

Herpes simplex virus (HSV) is from the *Herpes* viridae family and infects a high proportion of the populous. HSV-1 is generally responsible for cold sores and encephalitis, whereas HSV-2 is the main causative agent of anogenital infections, which can also infect neonates via the mother [59, 60]. Pomegranate rind extract (PRE) and its major ellagitannin compound, punicalagin, showed virucidal activity against HSV-1. While punicalagin has greater virucidal activity than an equivalent mass of pomegranate rind extract, PRE showed better antiviral activity than punicalagin. Moreover, PRE demonstrated comparable activity to acyclovir against HSV-1 and

Part of pomegranate used	Form	Test organisms	Mechanism of virus target	Reference
Rind	Crude hydraulic extract, Punicalagin, Ellagic acid	Herpes simplex virus	Virucidal activity	[48]
Juice, peel and pomegranate liquid extract		Influenze A viruses, H1N1, H3N2, H5N1 and coronavirus MHV A59	Damage to virion integrity and virucidal activity	[53]
Juice	Pomegranate polyphenol extract, punicalagin, pomegranate liquid extract (from POM Wonderful)	Human influenza A (H3N2)	Inhibition of viral RNA replication	[54]
Peel	Methanolic crude extract, punicalin, punicalagin, ellagic acid	Hepatitis C virus		[57]
Peel	Methanolic crude extract, ellagic acid, punicalagin, gallic acid	Adenovirus	Inhibition of adenovirus replication	[58]

Table 3.
Antiviral activity of pomegranate extracts against different viruses.

HSV-2, in addition to antiviral activity against acyclovir-resistant HSV-1 [48]. PRE is thus a promising new alternative treatment for HSV-1 since currently acyclovir is the gold standard treatment in HSV infections [61].

Studies have suggested that the antiviral activity of pomegranate extract originates from its hydrolysable tannins and polyphenols, especially punicalagin and gallagic acid. However, in one study, four flavonoids, ellagic acid, caffeic acid, luteolin and punicalagin, from pomegranate peel extract were studied against influenza virus and only punicalagin showed an inhibitory effect. The antiviral activity of pomegranate rind extract has been patented in Japan based on pomegranate peel extract ability to prevent the growth and kill viruses on the surfaces [46, 47]. The activity of pomegranate extract against viruses is summarized in **Table 3**.

5. Activity against parasites

Parasitic infections remain a significant global problem, affecting the health of hundreds of millions of people annually, especially in countries with low economic and social conditions. In addition, the increased world-wide resistance to conventional drugs is making most of currently used drugs less effective. As a result of this situation, the development of new drugs from medicinal plants for parasites is as important as for other microbes [62]. Different parts of *Punica granatum* L., root, stem bark, and rind of fruit, have been used commonly as vermifugal and taenicide agents [63]. The antiprotozoal activity of the pomegranate has been determined and in folkloric medicine, it has been used as anthelminthic especially against tapeworms and for diarrhea [64, 65]. A methanolic extract of pomegranate leaves showed nematicide activity and hepatoprotective activity against carbon tetrachloride induced hepatoxicity [66]. Extracts of pomegranate showed anti-schistosomal activity against *Shistosoma mansoni* in both *in vitro* and *in vivo* conditions [67].

Part of pomegranate used	Form	Mechanism of organism target	Organisms	Reference
Seed	Methanolic	Reduction in gastrointestinal motility		[65]
Leave	Methanolic	Larvicidal and ovicidal activity	*M. incognita, R. reniformis, P. penetrans, S. rolfsii*	[66]
Peels, juice and leaves	Methanolic	Reduction in viability of parasite	*Schistosoma mansoni, schistosomules*	[67]
Leaf, stem bark	Ethanolic	worms separation, reduction of motor activity, lethality, and ultrastructural tegumental alterations	*Schistosoma mansoni*	[68]
Peel	Powder form directly given to animal *in in vivo* study	Growth inhibition and death	*Cryptos poridium*	[72]
Juice	Crude extract was applied by patients in a clinical study		*Trichomoniasis vaginalis*	[73]

Table 4.
Antiparasitic activity of pomegranate extracts against different parasites.

In addition, it caused reduction or complete loss of motor activity, lethality and ultra-morphological changes in adult worms [68]. There is thus potential for the treatment of schistosomiasis.

Al-Musayeib *et al.* reported the antiparasitic activity of pomegranate rind extract against *Plasmodium falciparum* [69]. Pomegranate juice was found to exert dose-dependent activity against *Leishmania major* promastigotes and, at >80 μL/mL, gave significantly greater reduction than the positive control, Pentostam. Furthermore, mice that were orally treated with pomegranate juice, showed significantly reduced cutaneous leishmaniasis lesions compared to untreated mice [70]. Calzada *et al.* demonstrated pomegranate antiprotozoal activity against *Entameoba histolytica* and *Giardia lamblia* that cause diarrheic dysentery [71]. Pomegranate peel suspension also affected *C. parvum* in different stages and finally caused parasite death in an *in vivo* murine model; furthermore, pomegranate suspension did not cause any negative change in the mice ileal tissue [72]. In another study, pomegranate extract showed activity against *T. vaginalis*, both *in vitro* and clinically. Patients with *T. vaginalis* infection were treated with pomegranate juice and symptoms were found to have cleared after two months [73]. The activity of pomegranate extract against parasites is summarized in **Table 4**.

6. Potential mechanisms of antimicrobial activity of pomegranate extracts

From the preceding sections it is clear that there is compelling evidence demonstrating the broad-spectrum antimicrobial activity of pomegranate extracts [74–76]. However, the precise mechanism behind this activity is not fully understood. The mode of antimicrobial action of polyphenols, in general, is also unknown, although some suggested mechanisms include membrane disruption, toxicity against microorganisms, the ability of complex formation with metal ions and enzyme inactivation [77–79]. The antimicrobial activity of pomegranate has been associated with polyphenolic tannins, especially punicalagin and ellagic acid content in the extract [80–82]. However, pomegranate extracts are a complex mixture containing a variety of secondary compounds and interplay between these components may be a factor in antimicrobial activity, with multiple mechanisms operating independently [83].

An antimicrobial mechanism suggested for polyphenolic compounds is based on the precipitation ability of these compounds with bacterial cell membrane proteins which leads to bacterial cell lysis [84]. In addition, polyphenols could inhibit microbial enzymes by reacting with sulfhydryl groups or nonspecific interactions with proteins [85]. In that vein, phenolic compounds can bind the protein sulfhydryl groups and make them unavailable for microbial growth [86]. In addition, it has been reported that polyphenols can damage the microbe respiratory chain by decreasing the oxygen consumption and thus limiting the oxidation of NADH [87].

It has been hypothesized that the antibacterial activity of phenolic acids and flavonoids could cause a decrease in membrane fluidity by giving damage to the bacterial cytoplasmic membrane [88]. Phenolic acids can cause hyper acidification when they interphase with the plasma membrane. This situation would cause an alteration in cell membrane by making it more permeable. This mechanism could explain why phenolic acids show different antimicrobial activity levels against different pathogenic microorganisms [89, 90]. One of the possible mechanisms could be related to hydroxyl groups of polyphenols. The position of OH group in the aromatic ring and the length of saturated side chain could be a cause of antimicrobial activity of polyphenols [91]. Hydroxyl groups can bind to bacteria cell membranes and interfere with processes, such as ion pumping. In addition, OH groups can interact with active site of enzymes and disturb the metabolism of microorganisms [91].

Pomegranate extract exerted an inhibition activity against biofilms, in addition to their planktonic counterparts. Since microbes act differently under biofilm conditions compared to their planktonic form, there are some suggested pathways about polyphenols biofilm eradication and formation inhibition activities, although still unconfirmed. The mechanism behind growth and biofilm inhibition by pomegranate extracts cause protein precipitation and enzyme inactivation [81, 92]. Pomegranate extract could precipitate proteins which play major role in biofilm formation, like adhesins. Moreover, major hydrolysable tannins in pomegranate extract such as ellagic acid can change the surface charge and reduce the cell-substratum interactions and biofilm formation and development on different surfaces [93]. It is well known that tannins have astringency properties, and this feature can play a part in biofilm disruption [94, 95]. Different studies have shown the activity of pomegranate on bacterial attachment and therefore biofilm formation. It has been demonstrated that *Punica granatum* L. showed a specific antimicrobial action on dental plaque, which is a complex biofilm on tooth, by inhibiting adherence mechanism of oral microbes to dental surface via disturbing polyglucan synthesis [17, 96, 97]. Moreover, Vasconcelos *et al.* [98] used *Punica granatum* L. in a gel formulation using increasing and doubled concentrations of the diluted solutions of the gel with ranging concentrations from 1:1 to 1:1024, and similar results obtained. The gel formulation inhibited the adherence of different bacterial strains and a yeast, *C. albicans*, in the oral cavity and affected preformed biofilm.

There are some reports suggesting that the inhibition of quorum sensing (QS) could play role in the biofilm inhibition activity of pomegranate [99, 100]. QS is a communication system between bacteria in a biofilm, and provides a network involving nutrients, defense against other microorganisms, virulence and biofilm formation. More importantly, QS helps microbes to escape from host immune response [101, 102]. Therefore, inhibition of QS is quite important in order to overcome microbial infectious diseases and resistant pathogenic microbes. For the evaluation of QS inhibitors, *Chromobacterium violaceum* has been used as a biosensor since it produces violacein, purple pigment color, in response to QS regulation [103]. Pomegranate inhibited the QS of two bacterial strains which are *Chromobacterium violaceum* (by affecting purple pigment production) and *P. aeruginosa* (by decreasing bacterial swarming motility) [104, 105]. In another study, different compounds from herbs, fruits and plant extracts have been studied for their QS activity, with resveratrol and pomegranate extract demonstrating the highest inhibition activities. The QS activity of pomegranate has been associated with ellagic acid content of pomegranate extract (85% punicalagin, 7% free ellagic acid) since ellagic acid showed 86% inhibition at a low concentration of 4 µg/mL. However, the anti-QS activity of punicalagin is also believed important in pomegranate extracts [106]. Tannin-rich fraction of pomegranate rind extract showed inhibition of biofilm formation and motility of *E. coli* and repressed the expressions of curli genes (*csgB* and *csgD*) and various motility genes (*fimA, fimH, flhD, motB, qseB,* and *qseC*) [107]. Similarly, punicalagin significantly decreased the expression of QS-related genes (*sdiA* and *srgE*) of Salmonella [108].

The chemical structure of tannins has importance in bacterial growth inhibition. For example, hydrolysable tannins were found to give lower minimum inhibitory concentration than condensed tannins [109]. The degree of galloylation has an effect on antibacterial activity since a higher degree of galloylation have more protein binding capacity and higher affinity for iron. However, the antibacterial activity is not only correlated to galloyl groups and galloylation, also it is correlated to configuration of the digalloyl or trigalloyl groups that attached to glucose core [110–112]. In addition, free galloyl groups have a major role in antimicrobial activity of ellagitannins which are abundant secondary compounds in pomegranate extracts [12, 113]. Punicalagin showed the broad-spectrum antimicrobial activity and it has a gallagyl moiety [114]. However another ellgitannin, granatin A, which does not have free galloyl groups, did not show antibacterial activity [115]. In a study done by Reddy *et al.*, ellagic acid, gallagic acid, punicalin and punicalagin were purified and tested for their antiplasmodial and antimicrobial activities. Gallagic acid and punicalagin showed the strongest effects on the growth

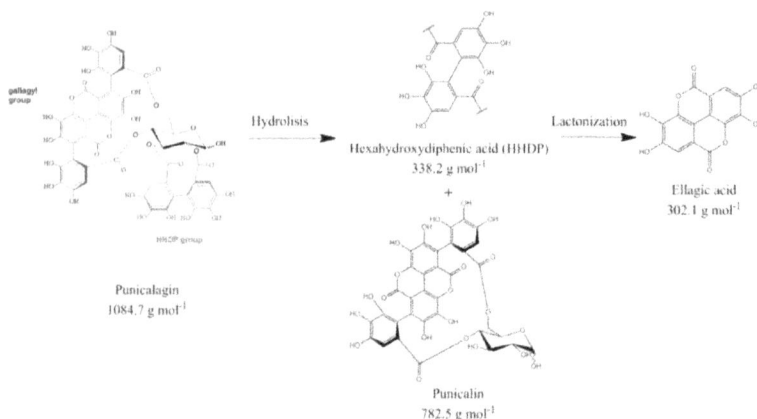

Figure 1.
Reduction of punicalagin, punicalin and HHDP to ellagic acid, adopted from Seeram et al. [3, 12].

of bacteria and fungi and it has been suggested that the ellagic acid moiety is not important compared to the gallagyl and hexahydroxydiphenol (HHDP) moieties for the inhibition of microbes [116]. The degradation of punicalagin to ellagic acid, via punicalin and hexahydroxydiphenic acid is shown in **Figure 1**.

The antimicrobial activity of plants has been studied extensively and many active secondary compounds have been identified. However, it should not be ignored that plant extracts with antimicrobial activities contain potentially many secondary compounds. Therefore, it is not easy to attribute the biological activity of plant extracts to only a single compound or a group of secondary compounds. There is a high possibility that plant extracts show antimicrobial activity due to synergistic effect of different compounds [117].

7. Enhanced antimicrobial activity of pomegranate extracts with metal ions

There are many reports showing the antimicrobial activity of heavy metals such as iron, copper, silver, manganese and zinc, while many bacteria have mechanism for the detoxification of heavy metals [118, 119]. However, although metal ions have a strong antimicrobial effect, they can also be cytotoxic to human cells, therefore, the use of these metals may have limitations in healthcare [120, 121].

Stewart *et al.* [122] investigated the potentiated antimicrobial activity of pomegranate rind extract (PRE) in combination with metal ions. In their study, the aim was to exert short term exposure of pomegranate rind extract and ferrous sulfate combination on bacteriophage levels for 3 minutes. This combination showed highly significant synergistic activity and reduced the bacteriophage levels in a short-term exposure. This short screening time was necessary for this experiment due to low stability of pomegranate rind extract/ferrous salt solution which, via a Fenton reaction caused Fe^{2+} to oxidize to Fe^{3+} with concomitant solution blackening. To overcome this instability problem, potentiated/synergised antimicrobial activity of pomegranate rind extract has since been examined using alternative metal ions [48, 123, 124].

McCarell *et al.* [123] investigated the antimicrobial activity of PRE with 4.6 mM $FeSO_4$, $CuSO_4$, $MnSO_4$, ZnO and also studied antimicrobial activity of PRE/metal salt combinations plus vitamin C which was added as a stabilizer. They observed significant synergistic antibacterial activity against *E. coli*, *Ps. Aeruginosa*, *S. aureus* and *P. mirabilis* when they combined PRE with Cu (II) ions. Moreover, with the addition of vitamin C as antioxidant, the antimicrobial activity increased significantly for PRE/Fe (II) and PRE/Cu (II) combinations against *S. aureus*. In another study, researchers used the vanillin complexes with different metal ions using the agar diffusion method and it was found that the vanillin and metal salts showed an enhanced activity against *S. aureus*, *E. coli*, *K. pneumanie*, *P. aeruginosa* and *C. albicans*. The results from both studies indicated that the addition of metal ions, especially copper salts, can significantly enhance antibacterial activity of a natural product [123, 125].

Significantly enhanced virucidal activity of PRE was later observed against HSV-1, HSV-2 and acyclovir-resistant HSV-1 by Houston *et al.* [48] in combination with different Zn (II) ion salts, including zinc sulphate, zinc citrate, zinc stearate and zinc gluconate, with a maximum of 6 log reduction observed. Unlike PRE and Fe^{2+}, this activity was not time-limited, and was not associated with blackening. Importantly, this activity was also retained when applied to epithelial surfaces prone to *Herpes* infection, including buccal and vaginal mucosae [126], indicating potential treatment for cold sores and anogenital *Herpes*.

The mechanism for the synergistic antimicrobial activity of pomegranate extract in combination with metal ions is not clear, although there are some putative suggested mechanisms for this enhanced antimicrobial activity. For instance, it has been suggested that pomegranate tannins can form a 'complex' with metallic ions and this complex could show enhanced toxicity to microbes [127]. Furthermore, PRE could show enhanced activity due to redox cycling of the associated metal ion which increases local levels of reactive oxygen species (ROS). For example some antibiotics e.g. bleomycin showed enhanced ROS production via the ability to bind to ferrous ions which resulted in enhanced toxicity against microbes [128].

The enhancement of antimicrobial activity of pomegranate rind extract with metal ions is important in terms of improved efficacy against antibiotic resistant pathogens, since this enhancement could reduce resistance of microbes by inhibiting their microbial adaptability features [8, 32].

8. Conclusions

The pomegranate has a long history of use as a folklore medicine for its ability to address microbial infections. Published research, as outlined in this chapter, clearly supports this and has shown that pomegranate extracts possess an unusual and potent broad-spectrum of activities against bacteria, fungi, viruses and parasites.

There is some variation in the literature in terms of the levels of antimicrobial activity of pomegranate extracts, which could be attributed to different harvesting time and type of pomegranate cultivars, and use of varying microbial strains. However, it is also apparent that different workers have used a range of approaches to obtain 'pomegranate extract', with extraction methods sometimes being poorly described. As such, a lack of standardized test extracts presents a challenge in attempting to make quantitative comparisons. As a complex mixture, pomegranates extracts have the innate ability to inhibit resistance, even more so when used alongside a synergizing metal ion.

Acknowledgements

We would like to thank to Turkish Ministry of Education for supporting Vildan Celiksoy's PhD project.

Conflict of interest

The authors declare no conflict of interest.

Author details

Vildan Celiksoy and Charles M. Heard*
School of Pharmacy and Pharmaceutical Sciences, Cardiff University,
United Kingdom

*Address all correspondence to: heard@cardiff.ac.uk

IntechOpen

References

[1] Bereket W, Hemalatha K, Getenet B, Wondwossen T, Solomon A, Zeynudin A, Kannan S. Update on bacterial nosocomial infections. Eur Rev Med Pharmacol Sci. 2012 Aug 1;16(8):1039-1044.

[2] Savard P, Perl TM. A call for action: managing the emergence of multidrug-resistant Enterobacteriaceae in the acute care settings. Current opinion in infectious diseases. 2012 Aug 1;25(4):371-7.

[3] Seeram NP, Adams LS, Henning SM, Niu Y, Zhang Y, Nair MG, Heber D. In vitro antiproliferative, apoptotic and antioxidant activities of punicalagin, ellagic acid and a total pomegranate tannin extract are enhanced in combination with other polyphenols as found in pomegranate juice. The Journal of nutritional biochemistry. 2005 Jun 1;16(6):360-7.

[4] Lansky EP, Newman RA. Punica granatum (pomegranate) and its potential for prevention and treatment of inflammation and cancer. Journal of ethnopharmacology. 2007 Jan 19;109(2):177-206.

[5] Braga LC, Shupp JW, Cummings C, Jett M, Takahashi JA, Carmo LS, Chartone-Souza E, Nascimento AM. Pomegranate extract inhibits Staphylococcus aureus growth and subsequent enterotoxin production. Journal of ethnopharmacology. 2005 Jan 4;96(1-2):335-9.

[6] Ahmad I, Beg AZ. Antimicrobial and phytochemical studies on 45 Indian medicinal plants against multi-drug resistant human pathogens. Journal of ethnopharmacology. 2001 Feb 1;74(2):113-23.

[7] Voravuthikunchai SP, Sririrak T, Limsuwan S, Supawita T, Iida T, Honda T. Inhibitory effects of active compounds from *Punica granatum* pericarp on verocytotoxin production by enterohemorrhagic Escherichia coli O157: H7. Journal of health science. 2005;51(5):590-6.

[8] Chidambara Murthy KN, Reddy VK, Veigas JM, Murthy UD. Study on wound healing activity of *Punica granatum* peel. Journal of Medicinal Food. 2004 Jun 1;7(2):256-9.

[9] Dey D, Debnath S, Hazra S, Ghosh S, Ray R, Hazra B. Pomegranate pericarp extract enhances the antibacterial activity of ciprofloxacin against extended-spectrum β-lactamase (ESBL) and metallo-β-lactamase (MBL) producing Gram-negative bacilli. Food and Chemical Toxicology. 2012 Dec 1;50(12):4302-9.

[10] Garcia-Villalba R, Espín JC, Aaby K, Alasalvar C, Heinonen M, Jacobs G, Voorspoels S, Koivumaki T, Kroon PA, Pelvan E, Saha S. Validated method for the characterization and quantification of extractable and nonextractable ellagitannins after acid hydrolysis in pomegranate fruits, juices, and extracts. Journal of agricultural and food chemistry. 2015 Jul 29;63(29):6555-66.

[11] Saad H, Charrier-El Bouhtoury F, Pizzi A, Rode K, Charrier B, Ayed N. Characterization of pomegranate peels tannin extractives. Industrial crops and Products. 2012 Nov 1; 40:239-46.

[12] Liu Y, Seeram NP. Liquid chromatography coupled with time-of-flight tandem mass spectrometry for comprehensive phenolic characterization of pomegranate fruit and flower extracts used as ingredients in botanical dietary supplements. Journal of separation science. 2018 Aug;41(15):3022-33.

[13] Zaouay F, Mena P, Garcia-Viguera C, Mars M. Antioxidant activity and

physico-chemical properties of Tunisian grown pomegranate (*Punica granatum* L.) cultivars. Industrial Crops and Products. 2012 Nov 1; 40:81-9.

[14] Panichayupakaranant P, Tewtrakul S, Yuenyongsawad S. Antibacterial, anti-inflammatory and anti-allergic activities of standardised pomegranate rind extract. Food Chemistry. 2010 Nov 15;123(2):400-3.

[15] Bakkiyaraj D, Nandhini JR, Malathy B, Pandian SK. The anti-biofilm potential of pomegranate (*Punica granatum* L.) extract against human bacterial and fungal pathogens. Biofouling. 2013 Sep 1;29(8):929-37.

[16] Avadhani M, Kukkamalla MA, Bhatt KG. Screening of *Punica granatum* extract for antimicrobial activity against oral microorganisms. Journal of Ayurvedic and Herbal Medicine. 2020;6(2):73-7.

[17] Benslimane S, Rebai O, Djibaoui R, Arabi A. Pomegranate Peel Extract Activities as Antioxidant and Antibiofilm against Bacteria Isolated from Caries and Supragingival Plaque. Jordan Journal of Biological Sciences. 2020 Jul 1;13(3).

[18] Sateriale D, Facchiano S, Colicchio R, Pagliuca C, Varricchio E, Paolucci M, Volpe MG, Salvatore P, Pagliarulo C. In vitro Synergy of Polyphenolic Extracts from Honey, Myrtle and Pomegranate Against Oral Pathogens, S. mutans and R. dentocariosa. Frontiers in Microbiology. 2020 Jul 24; 11:1465

[19] Rummun N, Somanah J, Ramsaha S, Bahorun T, Neergheen-Bhujun VS. Bioactivity of nonedible parts of *Punica granatum* L.: a potential source of functional ingredients. International journal of food science. 2013 Jul 8;2013.

[20] Vasconcelos LC, Sampaio FC, Sampaio MC, Pereira MD, Higino JS, Peixoto MH. Minimum inhibitory concentration of adherence of *Punica granatum* Linn (pomegranate) gel against S. mutans, S. mitis and C. albicans. Brazilian Dental Journal. 2006;17(3):223-7.

[21] Bhadbhade SJ, Acharya AB, Rodrigues SV, Thakur SL. The antiplaque efficacy of pomegranate mouthrinse. Quintessence International. 2011 Jan 1;42(1).

[22] DiSilvestro RA, DiSilvestro DJ, DiSilvestro DJ. Pomegranate extract mouth rinsing effects on saliva measures relevant to gingivitis risk. Phytotherapy Research: An International Journal Devoted to Pharmacological and Toxicological Evaluation of Natural Product Derivatives. 2009 Aug;23(8):1123-7.

[23] Martins ML, Ribeiro-Lages MB, Masterson D, Magno MB, Cavalcanti YW, Maia LC, Fonseca-Gonçalves A. Efficacy of natural antimicrobials derived from phenolic compounds in the control of biofilm in children and adolescents compared to synthetic antimicrobials: A systematic review and meta-analysis. Archives of Oral Biology. 2020 Jul 21:104844.

[24] Kanatt SR, Chander R, Sharma A. Antioxidant and antimicrobial activity of pomegranate peel extract improves the shelf life of chicken products. International journal of food science & technology. 2010 Feb 1;45(2):216-22.

[25] Negi PS, Jayaprakasha GK. Antioxidant, and antibacterial activities of *Punica granatum* peel extracts. Journal of food science. 2003 May;68(4):1473-7.

[26] Oliveira I, Sousa A, Morais JS, Ferreira IC, Bento A, Estevinho L, Pereira JA. Chemical composition, and antioxidant and antimicrobial activities of three hazelnut (*Corylus avellana*

L.) cultivars. Food and Chemical Toxicology. 2008 May 1;46(5):1801-7.

[27] Alpaslan D, Dudu TE, Şahiner N, Aktaşa N. Synthesis and preparation of responsive poly (Dimethyl acrylamide/gelatin and pomegranate extract) as a novel food packaging material. Materials Science and Engineering: C. 2020 Mar 1; 108:110339.

[28] Cui H, Surendhiran D, Li C, Lin L. Biodegradable zein active film containing chitosan nanoparticle encapsulated with pomegranate peel extract for food packaging. Food Packaging and Shelf Life. 2020 Jun 1; 24:100511.

[29] Loureiro MM, De Moraes BA, Mendonça VL, Quadra MR, Pinheiro GS, Asensi MD. Pseudomonas aeruginosa: study of antibiotic resistance and molecular typing in hospital infection cases in a neonatal intensive care unit from Rio de Janeiro City, Brazil. Memórias do Instituto Oswaldo Cruz. 2002 Apr;97(3):387-94.

[30] Morschhäuser J. The genetic basis of fluconazole resistance development in *Candida albicans*. Biochimica et Biophysica Acta (BBA)-Molecular Basis of Disease. 2002 Jul 18;1587(2-3):240-8.

[31] César de Souza Vasconcelos L, Sampaio MC, Sampaio FC, Higino JS. Use of *Punica granatum* as an antifungal agent against candidosis associated with denture stomatitis. Mycoses. 2003 Jun;46(5-6):192-6.

[32] Endo EH, Cortez DA, Ueda-Nakamura T, Nakamura CV, Dias Filho BP. Potent antifungal activity of extracts and pure compound isolated from pomegranate peels and synergism with fluconazole against *Candida albicans*. Research in Microbiology. 2010 Sep 1;161(7):534-40.

[33] Rizwan M, Mujtaba G, Memon SA, Lee K, Rashid N. Exploring the potential

of microalgae for new biotechnology applications and beyond: a review. Renewable and Sustainable Energy Reviews. 2018 Sep 1; 92:394-404.

[34] Costerton JW, Stewart PS, Greenberg EP. Bacterial biofilms: a common cause of persistent infections. Science. 1999 May 21;284(5418):1318-22.

[35] Komiyama EY, Mello de Matos B, Eduardo de Oliveira F, de Souza Reis T, Maynart de Faro H, Balducci I, Almeida JD, Koga-Ito CY. Proposal of Using Ozonated Water to Control Biofilm Formation on Mouth-Related Devices. Ozone: science & engineering. 2011 Sep 1;33(5):417-21.

[36] Hawser SP, Douglas LJ. Resistance of *Candida albicans* biofilms to antifungal agents in vitro. Antimicrobial agents and chemotherapy. 1995 Sep 1;39(9):2128-31.

[37] Sandasi M, Leonard CM, Van Vuuren SF, Viljoen AM. Peppermint (Mentha piperita) inhibits microbial biofilms in vitro. South African Journal of Botany. 2011 Jan 1;77(1):80-5.

[38] Endo EH, Ueda-Nakamura T, Nakamura CV. Activity of spray-dried microparticles containing pomegranate peel extract against *Candida albicans*. Molecules. 2012 Sep;17(9):10094-107.

[39] Heber D, Schulman RN, Seeram NP, editors. Pomegranates: ancient roots to modern medicine. CRC press; 2006 Jul 7

[40] Foss SR, Nakamura CV, Ueda-Nakamura T, Cortez DA, Endo EH, Dias Filho BP. Antifungal activity of pomegranate peel extract and isolated compound punicalagin against dermatophytes. Annals of clinical microbiology and antimicrobials. 2014 Dec 1;13(1):32.

[41] Salahvarzi Y, Tehranifar A, Jahanbakhsh V. Relation of antioxidant

and antifungal activity of different parts of pomegranate (*Punica granatum* L.) extracts with its phenolic content. Iranian Journal of Medicinal and Aromatic Plants. 2011;27(1).

[42] Ullah N, Ali J, Khan FA, Khurram M, Hussain A, Rahman IU, Rahman ZU, Ullah S. Proximate composition, minerals content, antibacterial and antifungal activity evaluation of pomegranate (*Punica granatum* L.) peels powder. Middle East J Sci Res. 2012;11(3):396-401.

[43] Glazer I, Masaphy S, Marciano P, Bar-Ilan I, Holland D, Kerem Z, Amir R. Partial identification of antifungal compounds from *Punica granatum* peel extracts. Journal of agricultural and food chemistry. 2012 May 16;60(19):4841-8.

[44] Negi PS. Plant extracts for the control of bacterial growth: Efficacy, stability and safety issues for food application. International journal of food microbiology. 2012 May 1;156(1):7-17.

[45] Viuda-Martos M, Ruiz Navajas Y, Sánchez Zapata E, Fernández-López J, Pérez-Álvarez JA. Antioxidant activity of essential oils of five spice plants widely used in a Mediterranean diet. Flavour and Fragrance Journal. 2010 Jan;25(1):13-9.

[46] Jassim SA, Denyer SP, Stewart GS, inventors; Merck Patent GmbH, assignee. Antiviral or antifungal composition comprising an extract of pomegranate rind or other plants and method of use. United States patent US 5,840,308. 1998 Nov 24.

[47] Jassim SA, Naji MA. Novel antiviral agents: a medicinal plant perspective. Journal of applied microbiology. 2003 Sep;95(3):412-27.

[48] Houston DM, Bugert JJ, Denyer SP, Heard CM. Correction: Potentiated

virucidal activity of pomegranate rind extract (PRE) and punicalagin against Herpes simplex virus (HSV) when co-administered with zinc (II) ions, and antiviral activity of PRE against HSV and aciclovir-resistant HSV. Plos one. 2017 Nov 20;12(11): e0188609.

[49] Sawai-Kuroda R, Kikuchi S, Shimizu YK, Sasaki Y, Kuroda K, Tanaka T, Yamamoto T, Sakurai K, Shimizu K. A polyphenol-rich extract from *Chaenomeles sinensis* (Chinese quince) inhibits influenza A virus infection by preventing primary transcription in vitro. Journal of ethnopharmacology. 2013 Apr 19;146(3):866-72.

[50] Das S, Tanwar J, Hameed S, Fatima Z, Manesar G. Antimicrobial potential of epigallocatechin-3-gallate (EGCG): a green tea polyphenol. J Biochem Pharmacol Res. 2014 Sep;2(3):167-74.

[51] Song JM, Lee KH, Seong BL. Antiviral effect of catechins in green tea on influenza virus. Antiviral research. 2005 Nov 1;68(2):66-74.

[52] Kamboj A, Saluja AK, Kumar M, Atri P. Antiviral activity of plant polyphenols. J Pharm Res. 2012 May;5(5):2402-12.

[53] Sundararajan A, Ganapathy R, Huan L, Dunlap JR, Webby RJ, Kotwal GJ, Sangster MY. Influenza virus variation in susceptibility to inactivation by pomegranate polyphenols is determined by envelope glycoproteins. Antiviral research. 2010 Oct 1;88(1):1-9.

[54] Haidari M, Ali M, Casscells III SW, Madjid M. Pomegranate (*Punica granatum*) purified polyphenol extract inhibits influenza virus and has a synergistic effect with oseltamivir. Phytomedicine. 2009 Dec 1;16(12):1127-36.

[55] Droebner K, Ehrhardt C, Poetter A, Ludwig S, Planz O. CYSTUS052, a

polyphenol-rich plant extract, exerts anti-influenza virus activity in mice. Antiviral research. 2007 Oct 1;76(1):1-0.

[56] Murzakhmetova M, Moldakarimov S, Tancheva L, Abarova S, Serkedjieva J. Antioxidant and prooxidant properties of a polyphenol-rich extract from *Geranium sanguineum* L. in vitro and in vivo. Phytotherapy Research: An International Journal Devoted to Pharmacological and Toxicological Evaluation of Natural Product Derivatives. 2008 Jun;22(6):746-51.

[57] Reddy BU, Mullick R, Kumar A, Sudha G, Srinivasan N, Das S. Small molecule inhibitors of HCV replication from pomegranate. Scientific reports. 2014 Jun 24; 4:5411.

[58] Karimi A, Moradi MT, Rabiei M, Alidadi S. In vitro anti-adenoviral activities of ethanol extract, fractions, and main phenolic compounds of pomegranate (*Punica granatum* L.) peel. Antiviral Chemistry and Chemotherapy. 2020 Apr; 28:2040206620916571.

[59] Whitley RJ, Nahmias AJ, Visintine AM, Fleming CL, Alford CA, Yeager A, Arvin A, Haynes R, Hilty M, Luby J. The natural history of herpes simplex virus infection of mother and newborn. Pediatrics. 1980 Oct 1;66(4):489-94.

[60] Whitley RJ, Roizman B. Herpes simplex virus infections. The lancet. 2001 May 12;357(9267):1513-8.

[61] Piret J, Boivin G. Resistance of herpes simplex viruses to nucleoside analogues: mechanisms, prevalence, and management. Antimicrobial agents and chemotherapy. 2011 Feb 1;55(2):459-72.

[62] Tagboto S, Townson S. Antiparasitic properties of medicinal plants and other naturally occurring products.

[63] Prakash CV, Prakash I. Bioactive chemical constituents from pomegranate (*Punica granatum*) juice, seed and peel-a review. International Journal of Research in Chemistry and Environment. 2011 Jul;1(1):1-8.

[64] Asres K, Bucar F, Knauder E, Yardley V, Kendrick H, Croft SL. In vitro antiprotozoal activity of extract and compounds from the stem bark of Combretum molle. Phytotherapy Research. 2001 Nov;15(7):613-7.

[65] Das AK, Mandal SC, Banerjee SK, Sinha S, Das J, Saha BP, Pal M. Studies on antidiarrhoeal activity of *Punica granatum* seed extract in rats. Journal of ethnopharmacology. 1999 Dec 15;68(1-3):205-8.

[66] Emam AM, Ahmed MA, Tammam MA, Hala AM, Zawam S. Isolation and structural identification of compounds with antioxidant, nematicidal and fungicidal activities from *Punica granatum* L. var. nana. International Journal of Scientific & Engineering Research. 2015;6(11):1023-40.

[67] Fahmy ZH, El-Shennawy AM, El-Komy W, Ali E, Hamid SA. Potential antiparasitic activity of pomegranate extracts against shistosomules and mature worms of Schistosoma Mansoni: in vitro and in vivo study. Australian Journal of Basic and Applied Sciences. 2009;3(4):4634-43.

[68] Yones DA, Badary DM, Sayed H, Bayoumi SA, Khalifa AA, El-Moghazy AM. Comparative evaluation of anthelmintic activity of edible and ornamental pomegranate ethanolic extracts against Schistosoma mansoni. BioMed research international. 2016 Jan 1;2016.

[69] Al-Musayeib NM, Mothana RA, Al-Massarani S, Matheeussen A, Cos P, Maes L. Study of the in vitro antiplasmodial, antileishmanial

and antitrypanosomal activities of medicinal plants from Saudi Arabia. Molecules. 2012 Oct;17(10):11379-90.

[70] Alkathiri B, El-Khadragy MF, Metwally DM, Al-Olayan EM, Bakhrebah MA, Abdel Moneim AE. Pomegranate (Punica granatum) juice shows antioxidant activity against cutaneous leishmaniasis-induced oxidative stress in female BALB/c mice. International Journal of Environmental Research and Public Health. 2017 Dec;14(12):1592.

[71] Calzada F, Yépez-Mulia L, Aguilar A. In vitro susceptibility of *Entamoeba histolytica* and *Giardia lamblia* to plants used in Mexican traditional medicine for the treatment of gastrointestinal disorders. Journal of Ethnopharmacology. 2006 Dec 6;108(3):367-70.

[72] Al-Mathal EM, Alsalem AA. Pomegranate (*Punica granatum*) peel is effective in a murine model of experimental *Cryptosporidium parvum* ultrastructural studies of the ileum. Experimental parasitology. 2013 Aug 1;134(4):482-94.

[73] El-Sherbini GM, Ibrahim KM, El-Sherbiny ET, Abdel-Hady NM, Morsy TA. Efficacy of *Punica granatum* extract on in-vitro and in-vivo control of Trichomonas vaginalis. Journal of the Egyptian Society of Parasitology. 2010;40(1):229-44.

[74] Thangavelu A, Elavarasu S, Sundaram R, Kumar T, Rajendran D, Prem F. Ancient seed for modern cure–pomegranate review of therapeutic applications in periodontics. Journal of pharmacy & bioallied sciences. 2017 Nov;9(Suppl 1): S11.

[75] Alexandre EM, Silva S, Santos SA, Silvestre AJ, Duarte MF, Saraiva JA, Pintado M. Antimicrobial activity of pomegranate peel extracts performed by high pressure and enzymatic assisted extraction. Food research international. 2019 Jan 1;115:167-76.

[76] Viana GS, Menezes SM, Cordeiro LN, Matos FJ. Biological Effects of Pomegranate (*Punica granatum* L.), especially its antibacterial actions, against microorganisms present in the dental plaque and other infectious processes. InBioactive Foods in Promoting Health 2010 Jan 1 (pp. 457-478). Academic Press.

[77] Papuc C, Goran GV, Predescu CN, Nicorescu V, Stefan G. Plant polyphenols as antioxidant and antibacterial agents for shelf-life extension of meat and meat products: Classification, structures, sources, and action mechanisms. Comprehensive Reviews in Food Science and Food Safety. 2017 Nov;16(6):1243-68.

[78] Bouarab Chibane L, Degraeve P, Ferhout H, Bouajila J, Oulahal N. Plant antimicrobial polyphenols as potential natural food preservatives. Journal of the Science of Food and Agriculture. 2019 Mar 15;99(4):1457-74.

[79] Daglia M. Polyphenols as antimicrobial agents. Current opinion in biotechnology. 2012 Apr 1;23(2):174-81.

[80] Al-Zoreky NS. Antimicrobial activity of pomegranate (*Punica granatum* L.) fruit peels. International journal of food microbiology. 2009 Sep 15;134(3):244-8.

[81] Fan W, Chi Y, Zhang S. The use of a tea polyphenol dip to extend the shelf life of silver carp (Hypophthalmicthys molitrix) during storage in ice. Food chemistry. 2008 May 1;108(1):148-53.

[82] Tehranifar A, Selahvarzi Y, Kharrazi M, Bakhsh VJ. High potential of agro-industrial by-products of pomegranate (*Punica granatum* L.) as the powerful antifungal and antioxidant substances. Industrial Crops and Products. 2011 Nov 1;34(3):1523-7.

[83] Bassole IH, Ouattara AS, Nebie R, Ouattara CA, Kabore ZI, Traore SA. Chemical composition and antibacterial activities of the essential oils of Lippia chevalieri and Lippia multiflora from Burkina Faso. Phytochemistry. 2003 Jan 1;62(2):209-12.

[84] Akhtar S, Ismail T, Fraternale D, Sestili P. Pomegranate peel and peel extracts: Chemistry and food features. Food chemistry. 2015 May 1; 174:417-25.

[85] Cowan MM. Plant products as antimicrobial agents. Clinical microbiology reviews. 1999 Oct 1;12(4):564-82.

[86] Haslam E. Natural polyphenols (vegetable tannins) as drugs: possible modes of action. Journal of natural products. 1996 Feb 22;59(2):205-15.

[87] Haraguchi H, Tanimoto K, Tamura Y, Mizutani K, Kinoshita T. Mode of antibacterial action of retrochalcones from Glycyrrhiza inflata. Phytochemistry. 1998 May 1; 48(1):125-9.

[88] Hatano T, Shintani Y, Aga Y, Shiota S, Tsuchiya T, Yoshida T. Phenolic constituents of licorice. VIII. Structures of glicophenone and glicoisoflavanone, and effects of licorice phenolics on methicillin-resistant Staphylococcus aureus. Chemical and Pharmaceutical Bulletin. 2000 Sep 1;48(9):1286-92.

[89] Miguel MG, Neves MA, Antunes MD. Pomegranate (*Punica granatum* L.): A medicinal plant with myriad biological properties-A short review. Journal of Medicinal Plants Research. 2010 Dec 29;4(25):2836-47.

[90] Lou Z, Wang H, Zhu S, Ma C, Wang Z. Antibacterial activity and mechanism of action of chlorogenic acid. Journal of food science. 2011 Aug;76(6):M398-403.

[91] Silva-Beltrán NP, Ruiz-Cruz S, Cira-Chávez LA, Estrada-Alvarado MI, Ornelas-Paz JD, López-Mata MA, Del-Toro-Sánchez CL, Ayala-Zavala JF, Márquez-Ríos E. Total phenolic, flavonoid, tomatine, and tomatidine contents and antioxidant and antimicrobial activities of extracts of tomato plant. International journal of analytical chemistry. 2015 Jan 1;2015.

[92] Naz S, Siddiqi R, Ahmad S, Rasool SA, Sayeed SA. Antibacterial activity directed isolation of compounds from *Punica granatum*. Journal of food science. 2007 Nov;72(9):M341-5.

[93] Lei Y, Tang Z, Liao R, Guo B. Hydrolysable tannin as environmentally friendly reducer and stabilizer for graphene oxide. Green chemistry. 2011;13(7):1655-8.

[94] Gregory WC, Gregory MP, Krapovickas A, Smith BW, Yarbrough JA. Structure and genetic resources of peanuts. Peanuts-culture and uses. 1973:47-133.

[95] Peng S, Jay-Allemand C. Use of antioxidants in extraction of tannins from walnut plants. Journal of chemical ecology. 1991 May 1;17(5):887-96.

[96] Ci Z, Kikuchi K, Hatsuzawa A, Nakai A, Jiang C, Itadani A, Kojima M. Antioxidant Activity, and α-Glucosidase, α-Amylase and Lipase Inhibitory Activity of Polyphenols in Flesh, Peel, Core and Seed from Mini Apple. American Journal of Food Science and Technology. 2018 Nov 15;6(6):258-62.

[97] Pereira JV, Pereira MS, Sampaio FC, Sampaio MC, Alves PM, Araújo CR, Higino JS. In vitro antibacterial and antiadherence effect of *Punica granatum* Linn extract upon dental biofilm microorganisms. Braz J Pharmacogn. 2006; 16:88-93.

[98] Vasconcelos LC, Sampaio FC, Sampaio MC, Pereira MD, Higino JS, Peixoto MH. Minimum inhibitory

concentration of adherence of *Punica granatum* Linn (pomegranate) gel against S. mutans, S. mitis and C. albicans. Brazilian Dental Journal. 2006;17(3):223-7.

[99] O'May C, Tufenkji N. The swarming motility of Pseudomonas aeruginosa is blocked by cranberry proanthocyanidins and other tannin-containing materials. Applied and environmental microbiology. 2011 May 1;77(9):3061-7.

[100] Sarabhai S, Sharma P, Capalash N. Ellagic acid derivatives from *Terminalia chebula* Retz. downregulate the expression of quorum sensing genes to attenuate Pseudomonas aeruginosa PAO1 virulence. PLoS one. 2013 Jan 8;8(1): e53441.

[101] Ni N, Li M, Wang J, Wang B. Inhibitors and antagonists of bacterial quorum sensing. Medicinal research reviews. 2009 Jan;29(1):65-124.

[102] Rudrappa T, Bais HP. Curcumin, a known phenolic from *Curcuma longa*, attenuates the virulence of Pseudomonas aeruginosa PAO1 in whole plant and animal pathogenicity models. Journal of Agricultural and Food Chemistry. 2008 Mar 26;56(6):1955-62.

[103] Morohoshi T, Kato M, Fukamachi K, Kato N, Ikeda T. N-acylhomoserine lactone regulates violacein production in Chromobacterium violaceum type strain ATCC 12472. FEMS microbiology letters. 2008 Feb 1;279(1):124-30.

[104] Koh KH, Tham FY. Screening of traditional Chinese medicinal plants for quorum-sensing inhibitors activity. Journal of Microbiology, Immunology and Infection. 2011 Apr 1;44(2):144-8.

[105] Zahin M, Hasan S, Aqil F, Khan M, Ahmad S, Husain FM,

Ahmad I. Screening of certain medicinal plants from India for their anti-quorum sensing activity.

[106] Truchado P, Tomás-Barberán FA, Larrosa M, Allende A. Food phytochemicals act as quorum sensing inhibitors reducing production and/or degrading autoinducers of Yersinia enterocolitica and Erwinia carotovora. Food Control. 2012 Mar 1;24(1-2):78-85.

[107] Yang Q, Wang L, Gao J, Liu X, Feng Y, Wu Q, Baloch AB, Cui L, Xia X. Tannin-rich fraction from pomegranate rind inhibits quorum sensing in Chromobacterium violaceum and biofilm formation in Escherichia coli. Foodborne pathogens and disease. 2016 Jan 1;13(1):28-35.

[108] Li G, Yan C, Xu Y, Feng Y, Wu Q, Lv X, Yang B, Wang X, Xia X. Punicalagin inhibits Salmonella virulence factors and has anti-quorum-sensing potential. Applied and environmental microbiology. 2014 Oct 1;80(19):6204-11.

[109] Ekambaram SP, Perumal SS, Balakrishnan A. Scope of hydrolysable tannins as possible antimicrobial agent. Phytotherapy Research. 2016 Jul;30(7):1035-45.

[110] Chung KT, Jr SS, Lin WF, Wei CI. Growth inhibition of selected food-borne bacteria by tannic acid, propyl gallate and related compounds. Letters in Applied Microbiology. 1993 Jul;17(1):29-32.

[111] Engels C, Gänzle MG, Schieber A. Fast LC–MS analysis of gallotannins from mango (*Mangifera indica* L.) kernels and effects of methanolysis on their antibacterial activity and iron binding capacity. Food research international. 2012 Jan 1;45(1):422-6.

[112] Tian F, Li B, Ji B, Zhang G, Luo Y. Identification and structure–activity

relationship of gallotannins separated from Galla chinensis. LWT-Food Science and Technology. 2009 Sep 1;42(7):1289-95.

[113] Farha AK, Yang QQ, Kim G, Li HB, Zhu F, Liu HY, Gan RY, Corke H. Tannins as an alternative to antibiotics. Food Bioscience. 2020 Sep 3:100751.

[114] Machado TD, Leal IC, Amaral AC, Santos K, Silva MG, Kuster RM. Antimicrobial ellagitannin of *Punica granatum* fruits. Journal of the Brazilian Chemical Society. 2002 Sep;13(5):606-10.

[115] Shimozu Y, Kimura Y, Esumi A, Aoyama H, Kuroda T, Sakagami H, Hatano T. Ellagitannins of Davidia involucrata. I. structure of davicratinic acid A and effects of davidia tannins on drug-resistant bacteria and human oral squamous cell carcinomas. Molecules. 2017 Mar;22(3):470.

[116] Reddy MK, Gupta SK, Jacob MR, Khan SI, Ferreira D. Antioxidant, antimalarial and antimicrobial activities of tannin-rich fractions, ellagitannins and phenolic acids from *Punica granatum* L. Planta medica. 2007 Oct;53(05):461-7.

[117] Compean KL, Ynalvez RA. Antimicrobial activity of plant secondary metabolites: A review. Research Journal of Medicinal Plants. 2014;8(5):204-13.

[118] Silver S. Bacterial resistances to toxic metal ions-a review. Gene. 1996 Jan 1;179(1):9-19.

[119] Ug A, Ceylan Ö. Occurrence of resistance to antibiotics, metals, and plasmids in clinical strains of Staphylococcus spp. Archives of Medical Research. 2003 Mar 1;34(2):130-6.

[120] O'Neill MA, Vine GJ, Beezer AE, Bishop AH, Hadgraft J, Labetoulle C,

Walker M, Bowler PG. Antimicrobial properties of silver-containing wound dressings: a microcalorimetric study. International journal of pharmaceutics. 2003 Sep 16;263(1-2):61-8.

[121] Strohal R, Schelling M, Takacs M, Jurecka W, Gruber U, Offner F. Nanocrystalline silver dressings as an efficient anti-MRSA barrier: a new solution to an increasing problem. Journal of Hospital Infection. 2005 Jul 1;60(3):226-30.

[122] Stewart GS, Jassim SA, Denyer SP, Newby P, Linley K, Dhir VK. The specific and sensitive detection of bacterial pathogens within 4 h using bacteriophage amplification. Journal of applied microbiology. 1998 May 1;84(5):777-83.

[123] McCarrell EM, Gould SW, Fielder MD, Kelly AF, El Sankary W, Naughton DP. Antimicrobial activities of pomegranate rind extracts: enhancement by addition of metal salts and vitamin C. BMC Complementary and Alternative Medicine. 2008 Dec;8(1):1-7.

[124] Houston D. Towards a nanomedicine-based broad-spectrum topical virucidal therapeutic system (Doctoral dissertation, Cardiff University).

[125] Nair MS, Joseyphus RS. Synthesis and characterization of Co (II), Ni (II), Cu (II) and Zn (II) complexes of tridentate Schiff base derived from vanillin and DL-α-aminobutyric acid. Spectrochimica Acta Part A: Molecular and Biomolecular Spectroscopy. 2008 Sep 1;70(4):749-53.

[126] Houston DM, Robins B, Bugert JJ, Denyer SP, Heard CM. In vitro permeation and biological activity of punicalagin and zinc (II) across skin and mucous membranes prone to Herpes simplex virus infection. European

Journal of Pharmaceutical Sciences.
2017 Jan 1; 96:99-106.

[127] Zhang L, Liu R, Gung BW,
Tindall S, Gonzalez JM, Halvorson JJ,
Hagerman AE. Polyphenol–aluminum
complex formation: implications for
aluminum tolerance in plants. Journal
of agricultural and food chemistry. 2016
Apr 20;64(15):3025-33

[128] Gould SW, Fielder MD,
Kelly AF, Sankary WE, Naughton DP.
Antimicrobial pomegranate rind
extracts: enhancement by Cu (II) and
vitamin C combinations against clinical
isolates of Pseudomonas aeruginosa.
British journal of biomedical science.
2009 Jan 1;66(3):129-32.

Chapter 3

Role of Pomegranate in the Management of Cancer

Amulya Thotambailu, Deepu Cheriamane,
Manjula Santhepete, Satheesh Kumar Bhandary,
Jiju Avanippully and Prakash Bhadravathi

Abstract

Pomegranate (*Punica granatum*) has been used since ages as a folk medicine. Studies have shown that pomegranate extracts have a role in various signaling pathways involved in inflammation, cellular transformation, hyperproliferation, angiogenesis, initiation of tumorigenesis, and eventually suppressing the final steps of tumorigenesis and metastasis. In this chapter, we have discussed some of the polyphenolic constituents present in pomegranate and their medical value, and we then discussed studies on chemopreventive/chemotherapeutic properties of pomegranate against various types of cancer, such as skin, prostate, colon, head and neck and lung cancers in cell culture systems, animal models and humans.

Keywords: Pomegranate, cancer, antioxidants

1. Introduction

The Pomegranate fruit (*Punica granatum* L.) is a perineal fruit of the family punicaceae, the fruit comprises of white to deep purple seeds covered in a spongy membrane which is further covered by the pericarp. It is available across the globe as a well-established folklore medicine owing to its anti-oxidant and anti-inflammatory properties. Various useful medicinally active components are present in the peel, seed, flower and even the leaves. The phenolic components comprise the major medicinally active part of the pomegranate extract along with minerals like magnesium, phosphorous, sodium and potassium [1–3]. There are four groups of phenolic compounds present in pomegranate namely the groups with anthocyanin pigments, hydrolysable tannins like punicalagin, ellagic acid and hydrolysable tannins. All of which contribute to the antioxidant activity. Moreover, it is rich in flavonoids and tannic acids which further adds to its the medicinal value. The antioxidant and anti-inflammatory potential of pomegranate juice and pomegranate extracts puts it apart from other fruits owing to the high concentrations of hydrolysable tannins and anthocyanins along with the polyphenols [4]. Among the various parts of the fruit, the peel and lamella which are the main on edible parts of the fruit has majority of the phenolic contents compared to the other edible parts. The peel accounts for more than half of the total antioxidative potential of the fruit and their anti-proliferative activity [5, 6]. The seed coat of the fruit also presents with numerous organic acids including citric acid

IntechOpen

and ascorbic acid [7]. Among the major biomedical advantages of pomegranate is the anti-cancer activity since pomegranate and its various components has been prove to assist the treatment of cancer and show immunomodulatory activity [8]. This chapter will be discussing on the anti-cancer activity of pomegranate and its antioxidant activity.

Cancer is one of the most common disease conditions which is becoming the leading cause of death even when detected in its early stages. In the year of 2021, almost 2 million new cancer cases are expected to happen just in the united states. The cancer death is reducing with each decade comparing to the initial few decades since its peak. With each passing year, there is rapid improvement in the cancer treatment strategies [9]. Pomegranate components can be used for treatment of many ailments as such or as an adjuvant in the treatment. One of the common problems related to cancer therapy is the lack of specificity in differentiating the cancer cells from the normal cells which manifests problems in the oral cavities as mucositis or candidiasis. This shows the prospect of using pomegranate extracts as an adjuvant in normal cancer chemotherapy in order to improve the quality of life of the people undergoing treatment. Also, the rind extract rich in the tannin punicalagin when used in combination with zinc is shows healing activity in the oral cavity due to the anti-inflammatory activity [10, 11].

2. Pomegranate and prostate cancer

Prostate cancer is the most prevalent type of cancer in men with an incidence rate above 30 worldwide [12]. From the multicenter studies conducted in human prostate cancer using pomegranate extracts rich in polyphenols, the extracts were found to cause an inhibition in the proliferation of the cells in both in *vivo* and *in vitro* studies. Hence the study demonstrated a significant anticancer activity through the inhibition of invasion and proliferation of the cancer cells [13]. The initial stages of prostate cancer will be testosterone dependent and this can be treated with normal radiation or chemotherapy but not in the late stages which shows no dependence with testosterone. The Further studies on the extracts from different parts of pomegranate showed a synergic activity in the anti-proliferative activity of other components of pomegranate. The seed oil from pomegranate acts as a synergistic when used in conjunction with the juice rich in polyphenols in the prevention of proliferation even though it does not have any anti-proliferative effect alone [14].

The studies on LNCaP cell lines which are modified to over express androgen receptors so that a situation similar to that of androgen independent prostate cancer. By using the different pomegranate extracts rich in polyphenols, the study later on showed a decrease in expression of the gene for the androgen synthesizing enzymes. Since the down regulation of androgen receptors is evident from the study, pomegranate extracts can be of use in the treatment of prostate cancer with an up-regulation of androgen receptors [15]. In androgen independent prostate cancer, there is an observed activation of the nuclear factor NF-κB. The activation of this nuclear factor is a common event in many types of cancer including breast cancer and cervical cancer [16]. In the molecular studies conducted using pomegranate extracts on the activity of NF-κB. It was found that the pomegranate extracts were able inhibit the NF-κB activity which was shown in the androgen independent cells, DU145 with increasing doses. Congruent results were obtained from the electro mobility shift assay conducted on the same cells using pomegranate extracts. In the DU145 and CL-1 cells which are the androgen independent cell lines, the activity of NF-κB was found to be activated through the TNF-α. Pomegranate extracts showed promising activity in the inhibition of NF-κB cells activated in this way as well. In the LAPC4 xenograft induced model of cancer, the extracts from pomegranate was found to delay the initiation of prostate cancer through prevention of proliferation of the cells [17].

Punicalagin is an important polyphenol constituent of pomegranate and as discussed before, the antioxidant activity of which is very evident in the cancer cells. The antiproliferative activity of punicalagin was examined in previous studies using the DPPH assay and the lipid peroxidation inhibition assays. Along with this, the study checked the cytotoxic activity and viability effects were also determined using punicalagin. It was found that punicalagin inhibited proliferation of cancer cells in prostate cancer and that the prostate cancer cells remained intact in the presence of punicalagin which was further supported by evidences from cell viability assays. The antioxidant activity of the polyphenol was further shown in the DPPH free radical scavenging assay which showed that it scavenged the free radicals in a dose dependent manner. The lipid peroxidation was also inhibited in the presence of punicalagin. PC-3 is another major cell line which is involved in prostate cancer and the polyphenol was found to reduce the PC-3 cells through apoptosis with higher concentrations [18, 19].

Further, it was found that pomegranate extracts affect the bio synthesis of androgens from the studies conducted using prostate cancer models. In the in vivo study conducted on the animal model using PTEN (Phosphatase and tensin homolog) knockout mouse which represents prostate cancer, there was observable reduction in the levels of steroids in the serum and in the case of in vitro studies using prostate cancer cell lines LNCaP and 22RV1, pomegranate extracts were found to cause a fall in the production of androgens. The in vitro and in vivo date obtained from various studies further shows the possible activity of pomegranate extracts in the treatment of Prostate cancer [20].

3. Pomegranate and breast cancer

Breast cancer is the most common type of cancer diagnosed in women and the leading cause of death due to cancer in women with over 2 million cases being diagnosed from recent studies [21]. The major causative factor for the cancer proliferation in breast cancer proliferation is estrogen and the enzymes which catalyzes the production of estrogen. The enzyme aromatase aids in the conversion of androgen into estrogen. So, the inhibition of this enzyme can further aid in the treatment of breast cancer. In vitro studies conducted on one of the major constituents of the pomegranate namely extract ellagic acid and urolithins A and B showed promising results on the inhibition of aromatase enzyme. The placental microsome aromatase assay conducted on ellagitannin derived compounds from pomegranate extracts namely, methylated urolithin B, methylated urolithin A and urolithin A further showed the aromatase inhibiting activity of pomegranate extracts. Which in turn inhibits the proliferation of cancer cells [22].

From the in vivo studies conducted on mammary organ culture in mice using the pomegranate seed oil rich in punicic acid and the fermented fruit extracts, it was found that the extracts of pomegranate caused a reduction in the number of lesions obtained and supports the activity of pomegranate extracts in the treatment of breast cancer [23].

The in vitro studies conducted on cancer stem cells derived from MMTV-Wnt-1, pomegranate extract was found to inhibit the proliferation of cancer cell by arresting the cell cycle at an early phase and induced apoptosis of the cancer cells. Pomegranate extracts caused an elevation I the levels of the enzyme caspase 3 which aids in the apoptosis. Among the various extracts, ellagic acid and ursolic acid along with luteolin were found to cause the inhibition of cell proliferation. Also, pomegranate extracts showed promising results in the molecular studies conducted on the MCF-7 cells of breast cancer through the inhibition of proliferation of the cancer cells. In the MCF-7 cells, the anti-cancer activity was found to be due to the cell cycle arrest, down regulation of genes which proliferate the cancer cells and also through the upregulation of the genes which aids in the regulation of proliferation and apoptosis. Hence, pomegranate extracts are relevant in the treatment of breast cancer therapy in the cases which are relatively resistant to the existing agents of treatment [24, 25].

4. Pomegranate and colon cancer

Colorectal cancer is currently one of the most common diagnosed cancer in men and women and it manifests with the uncontrolled proliferating of the epithelial cells and the suppression of their apoptosis [26]. One of the major constituents of pomegranate, the ellagitannin urolithin A plays a key role in the inhibition of proliferation of colon cancer cells through cell cycle arrest and the inhibition of mitogen activated protein kinase signaling (MAPK) [27].

The action of ellagitannins and urolithin on the CYP1 enzymes is important as these enzymes lead to the activation of inactive carcinogens into active carcinogenic chemicals in colon cancer. In the cell line study using HT-29 colon cancer cells, the evaluation of activity of CYP1 enzyme by employing EROD assay (ethoxy resorufin-*O*-deethylase assay) showed a reduction of CYP1 enzymes which were induced in the cell line. The extracts were found to show selective inhibition of proliferation of the cells omitting the non-cancer cells in a dose dependent fashion. Further, ellagitannin and urolithin was found to cause an increase in the apoptosis of the cell lines resulting in a reduction of the cell colony [28].

From the animal studies conducted on rats which were induced with colon cancer using N-methylnitrosourea which caused an increase in antigens which were specific to colon cancer along with and increase in plasma levels of Bcl2 and TGF-β, it was found that pomegranate peel extracts caused a fall in the cancer specific parameters which were induced in the mice. The in vivo study further suggests the efficacy of pomegranate in the treatment of colon cancer through the inhibition of proliferation and increased apoptosis which was evident from the fall in CEA and CCSA-4 prostate cancer cell markers along with the down regulation of β-catenin genes which has a pivotal role in the advancement of colon cancer. The down regulation of the specific gene disrupts the signaling pathway involving Wnt/β-catenin [29, 30].

From the cell line studies using HCT116 and HT-29 colon cancer cell lines, pomegranate extracts comprising of punicalagin, ellagic acid and tannins showed a drastic antiproliferative activity which led to complete inhibition of proliferation depending on the dose. The extract was found to cause apoptosis in the selected cell lines. Further, the extracts were found to have effect on the colon cancer cells which were not metastatic. The cell line studies further cement the role of punicalagin, ellagic acid and pomegranate tannins in the cancer protective activity in colon cancer [31].

5. Pomegranate and head and neck cancer

Head and neck cancers are one of the prevalent type of cancer which usually includes squamous cell carcinomas found in the epithelial cells of the pharynx, larynx and the oral cavity [32]. Due to the underdeveloped methods of screening of the disease, the chances of predicting the cancer at an early stage is less. This condition further leads to the increase in number of people who are diagnosed at a late stage of disease progression. In the current scenario, the treatment strategy of the disease mainly involves chemoradiation and surgery. The therapeutic approach to head and neck cancer comes with the common side effects of mucositis and dermatitis. Pomegranate extracts were studied for its protective effect in ameliorating the side effects of the treatment. In the clinical setup of a cohort containing patients with head and neck cancer, it was found that the extracts reduced the extend of damage caused by radiation induced dermatitis as well as mucositis [33].

Radiation therapy is applied in the cancer therapy for a long time because of its ability to kill the tumor cells but this will also lead to the production of reactive oxygen species that will damage the normal adjacent cells. Pomegranate extract has been studied in the amelioration of cellular damage induced by these reactive oxidants. From a study conducted using the extracts from pomegranate fruit and seeds, it was found that the treatment with the extracts increased the levels of antioxidant and the enzymes which has antioxidant property. Further the extracts were found to cause a decline in the lipid peroxidation levels suggesting the protective effect of pomegranate extracts in the cancer treatment as an adjuvant to reduce the unwanted side effects [34].

Other than in chemoradiation, pomegranate fruit extracts rich in punicalagin has been found useful in acting as a protective agent for the skin fibroblast cells namely the SKU-1064 from possible apoptosis due to UV-A and UV-B exposure. The extracts were found to suppress the NF- κB activation and through the downregulation of caspase-3 which is proapoptotic. Further studies found an increase in DNA repair through the increase in G0/G1 phase [35]. All these findings further supports the fact that pomegranate extracts can be applied in the treatment of cancer as an adjuvant also as a protective for radiation induced cellular damage.

6. Pomegranate and lung cancer

Lung cancer is one of the leading causes of death related to cancer worldwide in both men and women. Cigarette smoking is attributed to e the major cause of the condition. Along with lung cancer, cigarette smoke causes an increase in oxidative stress and DNA damage. From the animal studies conducted on the formation of lung nodules associated with lung cancer and other cancer related factors like the attenuation of mitosis and the levels of hypoxia inducible factor-1α or HIF1 α because of cigarette smoke, it was found that pomegranate juice supplements wee able to reduce the formation of lung nodules which is a common observation in the case of cigarette smoke exposure along with the reduction of mitosis and HIF-1 α [36].

Further, pomegranate fruit extract treatment in the cell line study using human carcinoma cell associated with lung cancer namely the A549 cells showed an inhibition of the markers of cell proliferation and angiogenesis such as MAPK, NIF-kappa B and PI3K/Akt. The treatment with the extracts further arrested the growth of tumor cells. Thus, pomegranate may be useful as a chemo preventive or as a chemotherapeutic agent against cancer affecting lungs [37]. Methotrexate is a widely used chemotherapeutic agent but it causes injuries in the lung cells due to oxidative stress. In the animal studies employed to study the effect of pomegranate extracts on the protective action against the lung injury caused due to methotrexate, it was found that the use of pomegranate extracts as a prophylactic significantly reduced the total oxidant status and the oxidative stress index along with elevating the total antioxidant capacity. This in turn shows the application of pomegranate extracts as an adjuvant as well in the therapy of lung cancer [38, 39].

7. Pomegranate and skin cancer

Pomegranate extracts from seed, peel and the whole fruit have been proven to be beneficial in the treatment of many cancer treatments. Skin cancer is the most common cancer among the Caucasian population and it varies depending on the type of cells affected. UV radiation is the major cause of skin cancer since it initiates and promotes tumor [40]. The in vivo and in vitro studies has shown the efficacy of pomegranate as a protectant in the UVB radiation induced skin damage. The oral treatment of pomegranate juice and extract in the Fitzpatrick II-IV skin type showed the possibility of enhancement in the protective from UV damage since it is able to increase the threshold of the UV dose required to cause erythema of skin [41].

The oil extracted from seed of pomegranate fruit was studied on animals as a topical prophylactic in the mice which was induced with skin cancer using 12-O-tetradecanoylphorbol 13-acetate and 7,12-dimethylbenzanthracene and it was found that the pre-treated animals had a significant less incidence of skin cancer which tells the protective effect of pomegranate in skin cancer [42]. From the studies conducted on the human skin fibroblasts SKU-1064 which were irradiated with UV, it was found that the pomegranate extracts rich in punicalagin was able to prevent the skin cell death showing the effect of pomegranate as a topical protective agent against skin cancer [35].

8. Conclusion

Pomegranate and the products derived from it has been proven to show various medicinal properties. Even though it has been in use in various traditional medical folklore since ages, the medicinal property of pomegranate is not explored much to be of use in the current medical scenario. Pomegranate is still being used as just a fruit and from the studies which are conducted so far on the fruit, it is to be noted that the extracts of the fruit rather than the whole fruit as such possesses many medicinal properties.

The role of pomegranate in the therapy of cancer as such and as an adjuvant in therapy is explored very less as there are very few studies has been conducted on humans, even though there are a handful of studies which are conducted on animal models or cell line studies which deems the fruit and its extracts effective in the therapy of cancer. The studies conducted so far shows the potency of pomegranate and its components in the treatment of cancer relating to prostate, breast, head and neck, colon, lungs and skin or as an adjuvant in the treatment to minimize the unwanted side effects. The various components of pomegranates because of its antioxidant and anti-inflammatory property can be applied to various treatment strategies in numerous types of cancer in one way or the other.

Hence it can be concluded that pomegranate extracts can be made to much use for humans in improving the treatment strategies in turn improving the quality of life, for which there has to be more human, animal and cell line studies so that the complete potency of pomegranate can be uncovered.

Author details

Amulya Thotambailu[1*], Deepu Cheriamane[2], Manjula Santhepete[3],
Satheesh Kumar Bhandary[4], Jiju Avanippully[5] and Prakash Bhadravathi[6]

1 Department of ENT, JSS Medical College and Hospital, JSS Academy of Higher
Education and Research, India

2 Department of Pulmonary Medicine, Sridevi Institute of Medical Sciences,
Tumkur, India

3 Department of Pharmacology, JSS College of Pharmacy, Mysuru, JSS Academy of
Higher Education and Research, Mysuru, India

4 ENT, NITTE (Deemed to be University), India

5 JSS College of Pharmacy, Mysuru, India

6 JSS Academy of Higher Education and Research, India

*Address all correspondence to: amulyathotambailu@gmail.com

IntechOpen

References

[1] Sharma J, Chandra R, Sharma K, Babu D, Meshram D, Maity A, et al. POMEGRANATE: Cultivation, Marketing and Utilization. 2014. https://doi.org/10.13140/RG.2.1.1286.4088.

[2] Naz S, Siddiqi R, Ahmad S, Rasool SA, Sayeed SA. Antibacterial Activity Directed Isolation of Compounds from Punica granatum. J Food Sci 2007;72:M341–M345. https://doi.org/10.1111/j.1750-3841.2007.00533.x.

[3] Viuda-Martos M, Fernández-López J, Pérez-Álvarez JA. Pomegranate and its Many Functional Components as Related to Human Health: A Review. Compr Rev Food Sci Food Saf 2010;9:635-654. https://doi.org/10.1111/j.1541-4337.2010.00131.x.

[4] Gil MI, Tomás-Barberán FA, Hess-Pierce B, Holcroft DM, Kader AA. Antioxidant Activity of Pomegranate Juice and Its Relationship with Phenolic Composition and Processing. J Agric Food Chem 2000;48:4581-4589. https://doi.org/10.1021/jf000404a.

[5] Orgil O, Schwartz E, Baruch L, Matityahu I, Mahajna J, Amir R. The antioxidative and anti-proliferative potential of non-edible organs of the pomegranate fruit and tree. LWT - Food Sci Technol 2014;58:571-577. https://doi.org/10.1016/j.lwt.2014.03.030.

[6] Mutahar S. S, Mutlag M. A-O, Najeeb S. A. Antioxidant Activity of Pomegranate (*Punica granatum* L.) Fruit Peels. Food Nutr Sci 2012;2012. https://doi.org/10.4236/fns.2012.37131.

[7] Viladomiu M, Hontecillas R, Lu P, Bassaganya-Riera J. Preventive and Prophylactic Mechanisms of Action of Pomegranate Bioactive Constituents. Evid Based Complement Alternat Med 2013;2013:789764. https://doi.org/10.1155/2013/789764.

[8] Xu Y-Y, Wang W-W, Huang J, Zhu W-G. Ellagic acid induces esophageal squamous cell carcinoma cell apoptosis by modulating SHP-1/STAT3 signaling. Kaohsiung J Med Sci 2020;36:699-704. https://onlinelibrary.wiley.com/doi/abs/10.1002/kjm2.12224 (accessed February 1, 2021).

[9] Siegel RL, Miller KD, Fuchs HE, Jemal A. Cancer Statistics, 2021. CA Cancer J Clin 2021;71:7-33. https://doi.org/10.3322/caac.21654.

[10] SANTOS MGC dos, NÓBREGA DR de M, ARNAUD RR, SANTOS RC dos, GOMES DQ de C, PEREIRA JV. Punica granatum Linn. prevention of oral candidiasis in patients undergoing anticancer treatment. Rev Odontol UNESP 2017;46:33-38.

[11] Celiksoy V, Moses RL, Sloan AJ, Moseley R, Heard CM. Evaluation of the In Vitro Oral Wound Healing Effects of Pomegranate (Punica granatum) Rind Extract and Punicalagin, in Combination with Zn (II). Biomolecules 2020;10:1234. https://doi.org/10.3390/biom10091234.

[12] Cancer today n.d. http://gco.iarc.fr/today/home (accessed February 5, 2021).

[13] Albrecht M, Jiang W, Kumi-Diaka J, Lansky EP, Gommersall LM, Patel A, et al. Pomegranate Extracts Potently Suppress Proliferation, Xenograft Growth, and Invasion of Human Prostate Cancer Cells. J Med Food 2004;7:274-283. https://doi.org/10.1089/jmf.2004.7.274.

[14] Lansky EP, Jiang W, Mo H, Bravo L, Froom P, Yu W, et al. Possible synergistic prostate cancer suppression by anatomically discrete pomegranate fractions. Invest New Drugs 2005;23:11-20. https://doi.org/10.1023/B:DRUG.0000047101.02178.07.

[15] Hong MY, Seeram NP, Heber D. Pomegranate polyphenols down-regulate expression of androgen-synthesizing genes in human prostate cancer cells overexpressing the androgen receptor. J Nutr Biochem 2008;19:848-855. https://doi.org/10.1016/j.jnutbio.2007.11.006.

[16] Baldwin AS Jr. Series Introduction: The transcription factor NF-κB and human disease. J Clin Invest 2001; 107:3-6. https://doi.org/10.1172/JCI11891.

[17] Rettig MB, Heber D, An J, Seeram NP, Rao JY, Liu H, et al. Pomegranate extract inhibits androgen-independent prostate cancer growth through a nuclear factor-κB-dependent mechanism. Mol Cancer Ther 2008;7:2662. https://doi.org/10.1158/1535-7163.MCT-08-0136.

[18] Adaramoye O, Erguen B, Nitzsche B, Höpfner M, Jung K, Rabien A. Punicalagin, a polyphenol from pomegranate fruit, induces growth inhibition and apoptosis in human PC-3 and LNCaP cells. Chem Biol Interact 2017;274:100-106. https://doi.org/10.1016/j.cbi.2017.07.009.

[19] Tai S, Sun Y, Squires JM, Zhang H, Oh WK, Liang C-Z, et al. PC3 is a cell line characteristic of prostatic small cell carcinoma. The Prostate 2011;71:1668-1679. https://doi.org/10.1002/pros.21383.

[20] Ming D-S, Pham S, Deb S, Chin MY, Kharmate G, Adomat H, et al. Pomegranate extracts impact the androgen biosynthesis pathways in prostate cancer models in vitro and in vivo. J Steroid Biochem Mol Biol 2014;143:19-28. https://doi.org/10.1016/j.jsbmb.2014.02.006.

[21] Bray F, Ferlay J, Soerjomataram I, Siegel RL, Torre LA, Jemal A. Global cancer statistics 2018: GLOBOCAN estimates of incidence and mortality worldwide for 36 cancers in 185 countries. CA Cancer J Clin 2018;68:394-424. https://doi.org/10.3322/caac.21492.

[22] Adams LS, Zhang Y, Seeram NP, Heber D, Chen S. Pomegranate Ellagitannin–Derived Compounds Exhibit Antiproliferative and Antiaromatase Activity in Breast Cancer Cells In vitro. Cancer Prev Res (Phila Pa) 2010;3:108-113. https://doi.org/10.1158/1940-6207.CAPR-08-0225.

[23] Mehta R, Lansky EP. Breast cancer chemopreventive properties of pomegranate (Punica granatum) fruit extracts in a mouse mammary organ culture. Eur J Cancer Prev 2004;13:345-348. https://doi.org/10.1097/01.cej.0000136571.70998.5a.

[24] Dai Z, Nair V, Khan M, Ciolino HP. Pomegranate extract inhibits the proliferation and viability of MMTV-Wnt-1 mouse mammary cancer stem cells in vitro. Oncol Rep 2010;24:1087-1091. https://doi.org/10.3892/or_00000959.

[25] Shirode AB, Kovvuru P, Chittur SV, Henning SM, Heber D, Reliene R. Antiproliferative effects of pomegranate extract in MCF-7 breast cancer cells are associated with reduced DNA repair gene expression and induction of double strand breaks. Mol Carcinog 2014;53:458-470. https://doi.org/10.1002/mc.21995.

[26] Zhao Y, Miao G, Li Y, Isaji T, Gu J, Li J, et al. Microrna 130b Suppresses Migration and Invasion of Colorectal Cancer Cells through Downregulation of Integrin β1. PLOS ONE 2014;9:1-10. https://doi.org/10.1371/journal.pone.0087938.

[27] González-Sarrías A, Espín J-C, Tomás-Barberán FA, García-Conesa M-T. Gene expression, cell cycle arrest

and MAPK signalling regulation in Caco-2 cells exposed to ellagic acid and its metabolites, urolithins. Mol Nutr Food Res 2009;53:686-698. https://doi.org/10.1002/mnfr.200800150.

[28] Kasimsetty SG, Bialonska D, Reddy MK, Ma G, Khan SI, Ferreira D. Colon Cancer Chemopreventive Activities of Pomegranate Ellagitannins and Urolithins. J Agric Food Chem 2010;58:2180-2187. https://doi.org/10.1021/jf903762h.

[29] Ahmed HH, El-Abhar HS, Hassanin EAK, Abdelkader NF, Shalaby MB. Punica granatum suppresses colon cancer through downregulation of Wnt/β-Catenin in rat model. Rev Bras Farmacogn 2017;27:627-635. https://doi.org/10.1016/j.bjp.2017.05.010.

[30] Ashihara E, Takada T, Maekawa T. Targeting the canonical Wnt/β-catenin pathway in hematological malignancies. Cancer Sci 2015;106:665-671. https://doi.org/10.1111/cas.12655.

[31] Seeram NP, Adams LS, Henning SM, Niu Y, Zhang Y, Nair MG, et al. In vitro antiproliferative, apoptotic and antioxidant activities of punicalagin, ellagic acid and a total pomegranate tannin extract are enhanced in combination with other polyphenols as found in pomegranate juice. J Nutr Biochem 2005;16:360-367. https://doi.org/10.1016/j.jnutbio.2005.01.006.

[32] Leemans CR, Braakhuis BJM, Brakenhoff RH. The molecular biology of head and neck cancer. Nat Rev Cancer 2011;11:9-22. https://doi.org/10.1038/nrc2982.

[33] Thotambailu AM, Bhandary BSK, Sharmila KP. Protective Effect of Punica granatum Extract in Head and Neck Cancer Patients Undergoing Radiotherapy. Indian J Otolaryngol Head Neck Surg 2019;71:318-320.

https://doi.org/10.1007/s12070-018-1297-4.

[34] Bhandary S, Sherly S, Kumari S, Bhat V, Sanjeev G. Ameliorative Activity of Punica granatum Extracts and Ellagic acid against Radiation Induced Biochemical Changes in Swiss Albino Mice. Res J Pharm Biol Chem Sci 2014;5:1097.

[35] Pacheco-Palencia LA, Noratto G, Hingorani L, Talcott ST, Mertens-Talcott SU. Protective Effects of Standardized Pomegranate (Punica granatum L.) Polyphenolic Extract in Ultraviolet-Irradiated Human Skin Fibroblasts. J Agric Food Chem 2008;56:8434-8441. https://doi.org/10.1021/jf8005307.

[36] Husari A, Hashem Y, Zaatari G, El Sabban M. Pomegranate Juice Prevents the Formation of Lung Nodules Secondary to Chronic Cigarette Smoke Exposure in an Animal Model. Oxid Med Cell Longev 2017;2017:e6063201. https://doi.org/10.1155/2017/6063201.

[37] Khan N, Mukhtar H. Pomegranate fruit as a lung cancer chemopreventive agent. Drugs Future - DRUG FUTURE 2007;32. https://doi.org/10.1358/dof.2007.032.06.1097137.

[38] Selimoğlu Sen H, Sen V, Bozkurt M, Türkçü G, Güzel A, Sezgi C, et al. Carvacrol and Pomegranate Extract in Treating Methotrexate-Induced Lung Oxidative Injury in Rats. Med Sci Monit Int Med J Exp Clin Res 2014;20:1983-1990. https://doi.org/10.12659/MSM.890972.

[39] Mukherjee S, Ghosh S, Choudhury S, Adhikary A, Manna K, Dey S, et al. Pomegranate reverses methotrexate-induced oxidative stress and apoptosis in hepatocytes by modulating Nrf2-NF-κB pathways. J Nutr Biochem 2013;24:2040-2050. https://doi.org/10.1016/j.jnutbio.2013.07.005.

[40] D'Orazio J, Jarrett S, Amaro-Ortiz A, Scott T. UV Radiation and the Skin. Int J Mol Sci 2013;14:12222-12248. https://doi.org/10.3390/ijms140612222.

[41] Henning SM, Yang J, Lee R-P, Huang J, Hsu M, Thames G, et al. Pomegranate Juice and Extract Consumption Increases the Resistance to UVB-induced Erythema and Changes the Skin Microbiome in Healthy Women: a Randomized Controlled Trial. Sci Rep 2019;9:14528. https://doi.org/10.1038/s41598-019-50926-2.

[42] Hora JJ, Maydew ER, Lansky EP, Dwivedi C. Chemopreventive Effects of Pomegranate Seed Oil on Skin Tumor Development in CD1 Mice. J Med Food 2003;6:157-161. https://doi.org/10.1089/10966200360716553.

Chapter 4

Vasculoprotective and Neuroprotective Effects of Various Parts of Pomegranate: In Vitro, In Vivo, and Preclinical Studies

Maria Trapali and Vasiliki Lagouri

Abstract

Pomegranate (*Punica granatum* L.) is one of the oldest edible fruits in the Mediterranean area and has been used extensively in the folk medicine. Popularity of pomegranate has increased especially in the last decade because of the health effects of the fruit. Polyphenols, represent the predominant class of phytochemicals of pomegranate, mainly consisting of hydrolysable tannins and ellagic acid. Pomegranate is a rich source of the ellagitannin punicalagin, which has aroused considerable interest in pomegranate fruit as a new therapeutic agent in recent years. Most studies on the effects of pomegranate juice have focused on its ability to cure diabetes and atherosclerosis. The present review summarizes some recent studies on the vasculoprotective and neuroprotective effect of various parts of pomegranate and its main compounds especially hydrolysable tannins ellagitannins, ellagic acid and their metabolites. The in vitro and in vivo studies, showed that the whole parts of pomegranate as well as its main components had a positive influence on blood glucose, lipid levels, oxidation stress and neuro/inflammatory biomarkers. They could be used as a future therapeutic agent towards several vascular and neurodegenerative disorders such as hypertension, coronary heart disease and Alzheimer.

Keywords: pomegranate, ellagic acid, punicalagin, urolithins, cardiovascular disease, CNS, in vitro, in vivo, pre-clinical trials

1. Introduction

Free radical reactions occur naturally in the human body. An over-production of these reactive species due to oxidative stress can cause oxidative damage to biomolecules and the development of chronic diseases such as aging, coronary heart disease and cancer [1]. The harmful action of free radicals can be inhibited by antioxidant substances which scavenge them and detoxify the organism. Current research has confirmed that dietary antioxidants play an important role in the prevention of cardiovascular diseases and cancers, neurodegenerative diseases and inflammation [2]. Pomegranate (*Punica granatum* L.) is one of the oldest edible fruits in the Mediterranean area and has been used extensively in the folk medicine. Popularity of pomegranate has increased in the last years because of anti-microbial, anti-viral, anti-cancer, anti-oxidant and anti-mutagenic effects of the fruit [3–5]. Polyphenols, are the main phytochemicals of pomegranate fruits, mainly consisting of hydrolysable tannins, gallotannins, ellagitannins and ellagic acid (EA). It has been found to exhibit antimutagenic, antiviral, whitening of the skin and antioxidative properties [6, 7]. Pomegranate fruit is composed of three different parts: the seeds, the arils and the peels. The therapeutic properties have been reported mostly for pomegranate juice [8–11] however, increasing literature was found lately reporting the inhibition of lipid peroxidation of pomegranate peels and seeds [4, 12, 13].

Even a small number of clinical trials in humans have been reported until now, the results showed positive effects of pomegranate extracts on various vascular diseases.

2. Phytochemical components related to activity

Ellagitannins (ETs) are esters of hexahydroxydiphenic acid (HHDP) and a polyol, usually glucose or quinic acid that when they are hydrolyzed transform through lactonization to the component ellagic acid [14] (**Figure 1**).

The variability in the chemical structures among ETs is associated with different physico-chemical properties, hydrolytic reactions, and biological activity in vivo [15]. The important structural diversity of ET structure is due to the different possible extent of galloylation and formation of aromatic C-glycosides, the number of intramolecular C-C coupling of galloyl groups and hydrolytic cleavage of galloyl-derived aromatic rings, the level of dehydrogenation, and oligomerization [16].

Ellagitannins and ellagic acid with anti-inflammatory and vasculoprotective effects are transformed by the gut microbiota to produce urolithins, bioavailable metabolites [17, 18] (**Figure 2**). There is, however, a large variability in health effects and can be associated with the different polyphenol glucuronide metabolic profiles. Differences in urolithin production, both quantity and chemical type, could explain, at least partly, the large variability in the health effects observed in vivo.

The effects of components of the pomegranate e.g. ellagic acid (EA) are also focusing on its potential protective action towards several neurodegenerative disorders. EA has been investigated as multi-target pharmacological drug on CNS in a review analysis [19]. Pomegranate metabolites such as urolithins prevented β-amyloid fibrillation in vitro and especially methyl-urolithin B (3-methoxy-6H-dibenzo [b, d] pyran-6-one), had a protective effect in *Caenorhabditis elegans* post induction of amyloid β(1–42) induced neurotoxicity and paralysis [20].

Urolithin A (UA) allayed hypoxia/reoxygenation abuse in myocardial cells, decreased myocardial cell death in mice after ischemia/reperfusion. UA enhanced antioxidant quantity in cardiomyocytes following hypoxia/reoxygenation reducing

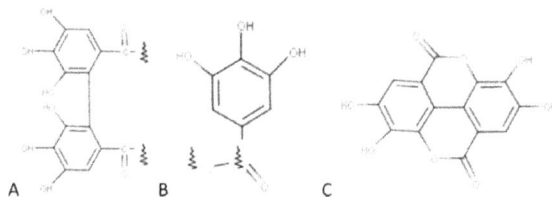

Figure 1.
Basic structures of ellagitannins: (A) HHDP acid (R radical); (B) galloyl unit (G radical); (C) ellagic acid.

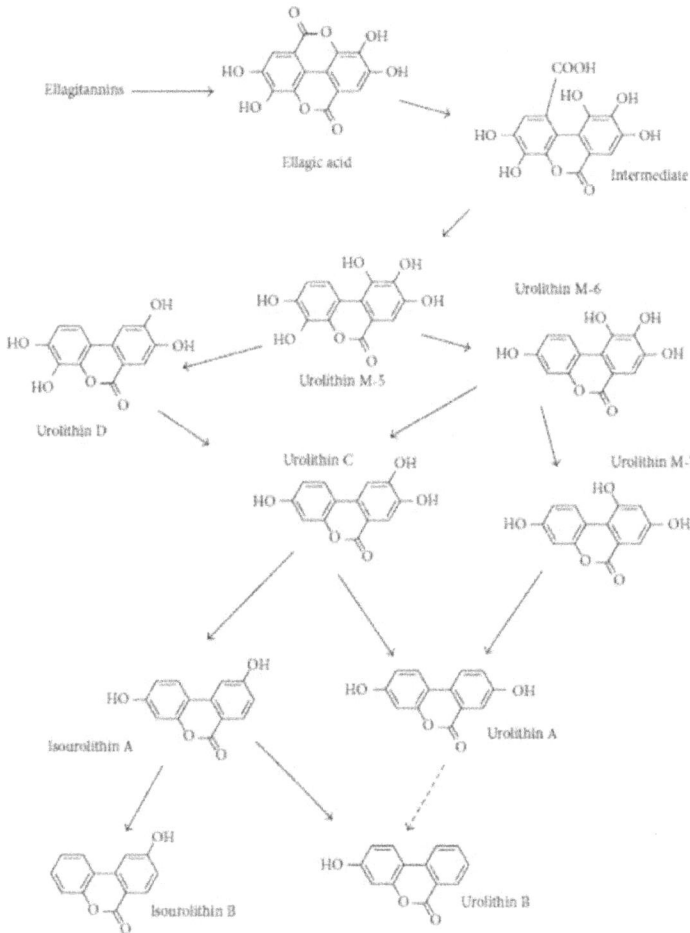

Figure 2.
Gut microbiota metabolism of ellagitannins and ellagic acid.

myocardial apoptosis [21]. The flavonoids naringin and narirutin have a significant beneficial effect in reducing diastolic blood pressure, in patients with hypertension [22]. Human umbilical vein endothelial cells (HUVECs) were pretreated
with ellagic acid and then incubated with oxidized low-density lipoprotein
(oxLDL). The results indicated inhibition of nicotinamide adenine dinucleotide
phosphate (NADPH) oxidase, enhancing cellular antioxidant defenses, and
attenuating oxLDL-induced Lectin-like oxidized low-density lipoprotein recep-
tor-1 (LOX-1) up-regulation and endothelial nitric oxide synthase (eNOS) downregulation. Lectin-like oxidized LDL (oxLDL) receptor-1 (LOX-1, also known
as OLR-1, is a class E scavenger receptor that mediates the uptake of oxLDL by
vascular cells. LOX-1 seems to represent an attractive therapeutic target for the
treatment of human atherosclerotic diseases [23]. Adipocyte cells were pretreated
with punicalagin and ellagic acid and that caused inhibition of lipolysis reducing
MAO activity [24].

Urolithin C, a combination of urolithins A and B metabolites of pomegranate
and ellagic acid also reduced cholesterol accumulation in the human monocytic cell
line THF-1-derived macrophages, but were unable to promote cholesterol outflow.
Atherosclerotic processes can be attenuated by urolithins, but future human

intervention tests are needed to see if it translates in vivo [23]. The ability of punicic acid (PUA) to modulate peroxisome proliferator-activated receptor PPAR activity was determined in 3 T3-L1 pre-adipocytes. PUA activates PPAR, increases PPAR - responsive gene expression and ameliorates diabetes and inflammation [25].

2.1 In vitro studies

PJ concentrate reduced the activation of redox-sensitive genes (ELK-1 and p-JUN) and increased eNOS expression in cultured human coronary artery endo-thelial cells (EC) exposed to high shear stress in vitro [26]. In vitro study showed that pomegranate leaf, seed and juice repressed cholinesterase activity in a dose dependent manner. Pomegranate juice had also protective effects against hydrogen peroxide induced toxicity in the *Artemia salina* (a species of brine shrimp) and HepG2 models (in vitro model system for the study of polarized human hepato-cytes), antiproliferative activities in HeLa and PC-3 cancer cells inhibiting COX-2 and MAO enzymes [27].

Microglial cells are the resident macrophages of the CNS. The immortalized murine microglial cell line BV-2 has been used frequently as a substitute for primary microglia. Urolithin B inhibited the production of NO and pro-inflammatory cyto-kines, inhibited NF-κB activity by reducing the phosphorylation and degradation of a nuclear factor of kappa light polypeptide gene enhancer in B-cells inhibitor, IκBα. In addition, urolithin B suppressed the phosphorylation of c-Jun N-terminal kinase (JNK), extracellular signal-regulated kinases (ERK), and Protein kinase B Akt, and enhanced the phosphorylation of AMPK, which is associated with anti-inflammatory and antioxidant processes [28, 29]. In another study, lipopolysac-charide LPS-treated cultured astrocytes and microglial BV-2 cells were investigated for anti-neuroinflammatory effects of punicalagin (PUN). It was found that PUN inhibits LPS-induced memory impairment via anti-inflammatory and anti-amylo-genic mechanisms through inhibition of nuclear factor kappa-light-chain-enhancer of activated B cells NF-κB activation [30]. The above results may be a solution to Alzheimer Disease [31].

2.2 In vivo studies

Clinical studies in hypertensive and/or obese patients receiving pomegranate juice have shown a reduction in systolic and diastolic blood pressure [32–36] and a concomitant increase in high density lipoprotein (HDL) cholesterol. Juice intake also led to a significant reduction in the by-products of fat peroxidation and protein and inflammatory biomarkers. Patients taking pomegranate-containing nutri-ent supplements had lowered systolic and diastolic blood pressure levels but the cardiovascular risk did not recover [37].

A number of clinical trials in humans proved the positive effects of pomegran-ate juice in the protection of central nervous system (CNS). Maternal pomegranate juice absorption in pregnancies with intrauterine growth restriction (IUGR) showed differences in the infant brain and structure [38].

2.3 Preclinical studies

When PJ was given in diabetic rats it was observed decreased blood glucose, lipid levels, and inflammatory biomarkers [39]. In another study using obese Zucker rats, intake of pomegranate juice (PJ) or fruit extract PFE caused a decrease of inflammation factors and increase of plasma nitrate and nitrite (NOx) [40]. In a study involving diabetic rats, they were given pomegranate seed powder (PS).

Pomegranate part/substance	Vasculoprotective effect (in vitro/in vivo)	Ref.	Neuroprotective effect (in vitro/in vivo)	Ref.
Pomegranate juice/peel extract/seed	Reduction in systolic and diastolic blood pressure (clinical studies/in vivo) Significant reduction in the by-products of fat peroxidation and protein and inflammatory biomarkers (clinical studies/in vivo) Decreased blood glucose, lipid levels, and inflammatory biomarkers (preclinical studies/in vivo) Improved insulin sensitivity, increased levels of interleukin-10 and activated PPARγ (preclinical studies/in vivo) Reduction of systemic oxidative stress (preclinical studies/in vivo) Reduced the activation of redox-sensitive genes (ELK-1 and p-JUN) and increased eNOS expression (in vitro)	Asgary et al. [32], Lynn et al. [35], Haghighian et al. [34], Asgary et al. [33], Moazzen and Alizadeh [36] Wu et al. [37] Taheri et al. [39], De Nigris et al. [40], Dos Santos et al. [42] Harzallah et al. [43], Hontecillas et al. [25] Asgary et al. [33] Nigris et al. [26]	Protection against oxidative destruction and improvement of neuronal durability (preclinical studies/in vivo) Inhibition of fetal brain apoptosis, neuronal nitric oxide synthase, and nuclear factor-κB activation (preclinical studies/in vivo) Repressed cholinesterase activity, Inhibition COX-2 and MAO-A enzymes (in vitro)	Kujawska et al. [44] Ginsberg et al. [45] Amri et al. [46], Les et al. [27]
Ellagic acid, punicalagin, urolithin	Inhibition of NADPH oxidase, enhancing cellular antioxidant defenses, attenuating oxLDL-induced LOX-1 up-regulation and eNOS down-regulation (in vitro) inhibition of lipolysis reducing MAO activity (in vitro) Increased PPAR-responsive gene expression and amelioration of diabetes and inflammation (in vitro)	Lee et al. [23] Les et al. [24] Hontecillas et al. [25] Yuan et al. [20]	Protective effect in neurotoxicity and paralysis (in vitro) Inhibition of the production of NO, pro-inflammatory cytokines, NF-κB activity, IκBα and Protein kinase B Akt (in vitro)	Yuan et al. [20] Lee et al. [29], DaSilva et al. [28], Kim et al. [30], AlMatar et al. [31]

Table 1.
Vasculoprotective and neuroprotective effects of pomegranate and their substances/metabolites in in vitro and in vivo pre-clinical studies.

Increased blood cholesterol, LDL and HDL lipoprotein were found [39, 41] while systolic blood pressure, angiotensin-converting enzyme coronary activity decreased [42]. Pomegranate peel (PPE), flower (PFE) and seed (PSO) given in obese mice decreased fasting blood glucose, improved insulin sensitivity, increased levels of the anti-inflammatory cytokine interleukin-10 [43] and activated peroxisome proliferator-activated receptor gamma (PPARγ) [25]. PPARγ, a ligand-activated transcription factor, has a role in various cellular functions as well as glucose homeostasis, lipid metabolism, and avoidance of oxidative stress. Pigs with hypercholesterolemia were given a pomegranate extract which caused reduction of systemic oxidative stress [33]. Pomegranate supplementation also exhibits cardiovascular protection improving cardiac hypertrophy in cigarette smoke in sight animals [11].

Preclinical trials in animal models added research results to the positive effects of pomegranate in CNS. In a rat model of Parkinsonism induced by rotenone, pomegranate juice treatment resulted in protection against oxidative destruction and improvement of neuronal durability [44]. Besides, in a rat model of maternal inflammation, pomegranate juice caused inhibition of fetal brain apoptosis, neuronal nitric oxide synthase, and nuclear factor-κB activation [45] (**Table 1**).

Methods used are extensively described in literature (e.g. [21, 37, 46–50]).

3. Possible therapeutic applications

The in vitro and in vivo studies showed that the whole parts of pomegranate as well as its main components such as hydrolysable tannins, ellagic acid and urolithins had a positive influence on blood glucose, lipid levels, oxidation stress and neuro/inflammatory biomarkers.

4. Future perspective and recommendations

The reviewed studies emphasize the potential benefits and suggest of a wider use of pomegranate and its components as dietary supplements or as adjuncts in the treatment of vascular and neurodegenerative diseases such as hypertension, coronary heart disease, peripheral artery disease and Alzheimer disease.

Conflict of interest

The authors declare that there are no conflicts of interest regarding the publication of this chapter.

Author details

Maria Trapali[1]* and Vasiliki Lagouri[2]

1 Laboratory of Chemistry, Biochemistry, Cosmetic Science, University of West Attica, Aigaleo, Greece

2 Institute of Chemical Biology, National Hellenic Research Foundation, Athens, Greece

*Address all correspondence to: mariatrapali66@yahoo.gr

IntechOpen

References

[1] Halliwell, B.; Gutteridge, J.M.C. and Cross, C.E. (1992). Free radicals, antioxidants and human disease: where are now? J. Lab. Clin. Med. 119, 598-619

[2] Scalbert, A.; Manach, C.; Morand, C. and Remesy, C. (2005). Dietary of polyphenols and the prevention of diseases. Crit. Rev. Food Sci. Nutr. 45, 287-30651

[3] Lansky, E. P., & Newman, R. A. (2007). Punica granatum (pomegranate) and its potential for prevention and treatment of inflammation and cancer. Journal of Ethnopharmacology, 109, 177-206.

[4] Li, Y., Guo, C., Yang, J., Wei, J., Xu, J., & Cheng, S. (2006). Evaluation of antioxidant properties of pomegranate peel extract in comparison with pomegranate pulp extract. Food Chemistry. 96, 254-260.

[5] Negi, P. S., Jayaprakasha, G. K., & Jena, B. S. (2003). Antioxidant and antimutagenic activities of pomegranate peel extracts. Food Chemistry. 80, 393-397.

[6] Seeram, N. P.; Adamsa, L. S.; Henninga, S. M.; Niu, Y.; Zhang, Y.; Nair, M. G. and Heber, D. (2005). In vitro antiproliferative, apoptotic and antioxidant activities of punicalagin, ellagic acid and a total pomegranate tannin pomegranate juice. J. of Nutritional Biochemistry, Vol.16, Iss.6, 360-67

[7] Vattem D.A. and Shetty K. (2005). Biological functionality of ellagic acid: A review. J. of Food Biochemistry, Vol.29, Iss.3, 234-66

[8] Gil, M. I.; Tomas B., F. A.; Hess P., B.; Holcroft, D. M.; Kader, A. A., (2000). Antioxidant activity of pomegranate juice and its relationship with phenolic composition and processing. J. Agric. Food Chem (48), 4581-4589

[9] Tezcan F., Gültekin Ö. M., Diken T., Özçelik B., Bedia E. F., (2009). Antioxidant activity and total phenolic, organic acid and sugar content in commercial pomegranate juices. Food Chemistry 115, 873-877

[10] Tzulker R., Glazer I., Bar I. I., Holland D., Aviram M., & Amir R., (2007). Antioxidant Activity, Polyphenol Content, and Related Compounds in Different Fruit Juices and Homogenates Prepared from 29 Different Pomegranate Accessions. J. Agric. Food Chem. 55, 9559-9570.

[11] Wang D., Özen C., Abu-ReidahI., Chigurupati S., Patra J., Horbanczuk J., Józ'wik A., Tzvetkov N., Uhrin P., Atanasov A. (2018) Vasculoprotective Effects of Pomegranate (*Punica granatum* L.), Frontiers in Pharmacology; 9, 544. doi: 10.3389/fphar.2018.00544

[12] Singh R. P., Chidambara M. K. N., & Jayaprakasha G. K., (2002). Studies on the Antioxidant Activity of Pomegranate (Punica granatum) Peel and Seed Extracts Using in Vitro Models. J. Agric. Food Chem. 50, 81-86

[13] Çam M Yaşar H (2010). Pressurised water extraction of polyphenols from pomegranate peels. Food Chem 123, 878-885

[14] Koponen J.M., Happonen A.M., Mattila P.H., Torronen A.R. (2007). Contents of anthocyanins and ellagitannins in selected foods consumed in Finland. Journal of Agricultural and Food Chemistry, 55:1612-1619

[15] Lipińska L., Klewicka E., Sójka M. (2014). Structure, occurrence and biological activity of ellagitannins: A general review. Acta Scientiarum Polonorum. Technologia Alimentaria. 13(3):289-299

[16] Yamada H, Wakamori S. (2018). Structural revisions in natural ellagitannins. Molecules. 23:1-46

[17] Espín, J. C., Larrosa, M., García-Conesa, M. T., Tomás-Barberán, F. A. (2013). Biological significance of urolithins, the gut microbial ellagic acid- derived metabolites: the evidence so far. Evidence-Based Complement. Altern. Med. DOI:10.1155/2013/270418

[18] Larrosa, M., García-Conesa, M. T., Espín, J. C., Tomás-Barberán, F. A. (2010). Ellagitannins, ellagic acid and vascular health. Mol. Aspects Med. 31, 513-539.

[19] Alfei S., Turrini F., Catena S., Zunin P., Grilli M., Pittaluga A., Boggia R. (2019) Ellagic acid a multi-target bioactive compound for drug discovery in CNS? A narrative review. Eur J Med Chem. 183, 11724 doi: 10.1016/j.ejmech.2019.111724.

[20] Yuan T., Ma H., Liu W., Niesen D., Shah N., Crews R., Rose K., Vattem D., Seeram N. (2016) Pomegranate's Neuroprotective Effects against Alzheimer's Disease Are Mediated by Urolithins, Its Ellagitannin-Gut Microbial Derived Metabolites. ACS Chem Neurosci. 7(1),26-33. doi: 10.1021/acschemneuro.5b00260.

[21] Tang L., Mo Y., Li Y., Zhong Y., He S., Zhang Y., Tang Y., Fu S., Wang X., Chen A. (2017) Urolithin A alleviates myocardial ischemia/reperfusion injury via PI3K/Akt pathway. Biochem Biophys Res Commun. 486(3),774-780. doi: 10.1016/j.bbrc.2017.03.119

[22] Reshef N., Hayari Y., Goren C., Boaz M. (2005) Antihypertensive Effect of Sweetie Fruit in Patients with Stage I Hypertension. American Journal of Hypertension 18(10), 1360-1363 DOI: 10.1016/j.amjhyper.2005.05.021

[23] Lee, W. J., Ou, H. C., Hsu, W. C., Chou, M. M., Tseng, J. J., Hsu, S. L., et al. (2010). Ellagic acid inhibits oxidized LDL-mediated LOX-1 expression, ROS generation, and inflammation in human endothelial cells. J. Vasc. Surg. 52, 1290-1300. doi: 10.1016/j.jvs.2010.04.085

[24] Les, F., Carpene, C., Arbones-Mainar, J. M., Decaunes, P., Valero, M. S., and Lopez, V. (2017). Pomegranate juice and its main polyphenols exhibit direct effects on amine oxidases from human adipose tissue and inhibit lipid metabolism in adipocytes. J. Funct. Foods 33, 323-331. doi: 10.1016/j. jff.2017.04.006

[25] Hontecillas R., O'Shea M., Einerhand A., Diguardo M., Bassaganya-Riera J. Activation of PPAR g and a by punicic acid ameliorates glucose tolerance and suppresses obesity-related inflammation. J. Am. Coll. Nutr. 2009,28: 184-195. doi: 10.1080/07315724.2009.10719770.

[26] Nigris F., Williams-Ignarro S., Lerman L., Crimi E., Botti C., Mansueto G., D'Armiento F., De Rosa G., Sica V., Ignarro L., Napoli C. (2005) Beneficial effects of pomegranate juice on oxidation-sensitive genes and endothelial nitric oxide synthase activity at sites of perturbed shear stress. PNAS.102 (13), 4896-4901 https://doi. org/10.1073/pnas.0500998102

[27] Les F., Prieto J. M., Arbonés-Mainar J. M., Valero M. S., López V. (2015). Bioactive properties of commercialised pomegranate (Punica granatum) juice: antioxidant, antiproliferative and enzyme inhibiting activities. Food Funct. 6, 2049-2057. 10.1039

[28] DaSilva N., Nahar P., Ma H., Eid A., Wei Z., Meschwitz S., Zawia N., Slitt A., Seeram N. (2019) Pomegranate ellagitannin-gut microbial-derived metabolites, urolithins, inhibit neuroinflammation in vitro. Nutr. Neurosci.22(3),185-195 doi: 10.1080/1028415X.2017.1360558.

[29] Lee G., Park J., Jung Lee E., Ahn J., Kim H. (2019) Anti-inflammatory and antioxidant mechanisms of urolithin B in activated microglia. Phytomedicine 55, 50-57 doi: 10.1016/j.phymed.2018.06.032

[30] Kim Y.., Hwang C., Lee H., Kim C., Son D., Ham Y., Hellström M., Han S., Kim H. , Park E., Hong J. (2017) Inhibitory effect of punicalagin on lipopolysaccharide-induced neuroinflammation, oxidative stress and memory impairment via inhibition of nuclear factor-kappaB. Neuropharmacology 117, 21-32 doi: 10.1016/j.neuropharm.2017.01.025.

[31] AlMatar M., Islam M., Albarri O., Var I., Koksal F. (2018) Risk of Developing Wound Healing, Obesity, Neurodegenerative Disorders, and Diabetes Mellitus. Mini Rev Med Chem. 18(6),507-526 doi: 10.2174/13895575176 66170419114722.

[32] Asgary S., Keshvari M., Sahebkar A., Hashemi M., Rafieian-Kopaei M. (2013). Clinical investigation of the acute effects of pomegranate juice on blood pressure and endothelial function in hypertensive individuals. ARYA Atheroscler. 9,326-331

[33] Asgary, S., Keshvari, M., Sahebkar, A., and Sarrafzadegan, N. (2017). Pomegranate consumption and blood pressure: a review. Curr. Pharm. Des. 23, 1042-1050. doi: 10.2174/13816128226661 61010103339

[34] Haghighian, M. K., Rafraf, M., Moghaddam, A., Hemmati, S., Jafarabadi, M. A., and Gargari, B. P. (2016). Pomegranate (*Punica granatum* L.) peel hydro alcoholic extract ameliorates cardiovascular risk factors in obese women with dyslipidemia: a double blind, randomized, placebo-controlled pilot study. Eur. J. Integr. Med. 8, 676-682. doi: 10.1016/j.eujim.2016.06.010

[35] Lynn, A., Hamadeh, H., Leung, W. C., Russell, J. M., and Barker, M. E. (2012). Effects of pomegranate juice supplementation on pulse wave velocity and blood pressure in healthy young and middle-aged men and women. Plant Foods Hum. Nutr. 67, 309-314. doi: 10.1007/s11130-012-0295-z

[36] Moazzen H., Alizadeh M. (2017) Effects of pomegranate juice on cardiovascular risk factors in patients with metabolic syndrome: A double-blinded, randomized cross over controlled trial. Plant Foods Hum. Nutr. 72,126-133. doi: 10.1007/s11130-017-0605-6.

[37] Wu P.T., Fitschen P.J., Kistler B.M., Jeong J.H., Chung H.R., Aviram M., Phillips S.A., Fernhall B., Wilund K.R. (2015) Effects of pomegranate extract supplementation on cardiovascular risk factors and physical function in hemodialysis patients. J. Med. Food.18,941-949 doi: 10.1089/jmf.2014.0103.

[38] Matthews L., Smyser C., Cherkerzian S., Alexopoulos D., Kenley J., Tuuli M., Nelson D., Inder T. (2019) Maternal pomegranate juice intake and brain structure and function in infants with intrauterine growth restriction: A randomized controlled pilot study. PLoS One. 14(8), e0219596.

[39] Taheri R., Sarker M.M., Rahmat A., Alkahtani S.A., Othman F. (2017) The effect of pomegranate fresh juice versus pomegranate seed powder on metabolic indices, lipid profile, inflammatory biomarkers, and the histopathology of pancreatic islets of Langerhans in streptozotocin-nicotinamide induced type 2 diabetic Sprague-Dawley rats. BMC Complement. Altern. Med. 17 doi: 10.1186/s12906-017-1667-6

[40] De Nigris F., Balestrieri M.L., Williams-Ignarro S., D'Armiento F.P., Fiorito C., Ignarro L.J., Napoli C.

(2007) The influence of pomegranate fruit extract in comparison to regular pomegranate juice and seed oil on nitric oxide and arterial function in obese Zucker rats. Nitric Oxide. 17,50-54. doi: 10.1016/j.niox.2007.04.005.

[41] Huang, T. H., Peng, G., Kota, B. P., Li, G. Q., Yamahara, J., Roufogalis, B. D., et al. (2005). Anti-diabetic action of Punica granatum flower extract: activation of PPAR-gamma and identification of an active component. Toxicol. Appl. Pharmacol. 207, 160-169. doi: 10.1016/j.taap.2004.12.009

[42] Dos Santos, R. L., Dellacqua, L. O., Delgado, N. T., Rouver, W. N., Podratz, P. L., Lima, L. C., et al. (2016). Pomegranate peel extract attenuates oxidative stress by decreasing coronary angiotensin-converting enzyme (ACE) activity in hypertensive female rats. J. Toxicol. Environ. Health A 79, 998-1007. doi: 10.1080/15287394.2016.1213690

[43] Harzallah A., Hammami M., Kępczyńska M.A., Hislop D.C., Arch J.R., Cawthorne M.A., Zaibi M.S. (2016) Comparison of potential preventive effects of pomegranate flower, peel and seed oil on insulin resistance and inflammation in high-fat and high-sucrose diet-induced obesity mice model. Arch. Physiol. Biochem. 122, 75-87. doi: 10.3109/13813455.2016.1148053.

[44] Kujawska M., Jourdes M., Kurpik M., Szulc M., Szaefer H., Chmielarz P., Kreiner G., Krajka-Kuźniak V., Mikołajczak P., Teissedre P., Jodynis-Liebert J. (2020) Neuroprotective Effects of Pomegranate Juice against Parkinson's Disease and Presence of Ellagitannins-Derived Metabolite—Urolithin A—In the Brain. Int J Mol Sci. 21(1), 202.

[45] Ginsberg Y., Khatib N., Saadi N., Ross M., Weiner Z., Beloosesky R. (2018) Maternal pomegranate juice attenuates maternal inflammation-induced fetal brain injury by inhibition of apoptosis, neuronal nitric oxide synthase, and NF-κB in a rat model. Am J Obstet Gynecol 219(1),113e1-113.e9. doi: 10.1016/j.ajog.2018.04.040.

[46] Amri Z., Ghorbel A., Turki M., Akrout M., Ayadi F., Elfeki A., Hammami M. (2017) Effect of pomegranate extracts on brain antioxidant markers and cholinesterase activity in high fat-high fructose diet induced obesity in rat model. BMC Complement Altern Med., 17, 339.

[47] Benzie IF, Strain JJ (1996) The ferric reducing ability of plasma (FRAP) as a measure of "antioxidant power": The FRAP assay. Anal Biochem 239,70-76

[48] Kojadinovic M.I., Arsic A.C., Debeljak-Martacic J.D., Konic-Ristic A.I., Kardum N.D., Popovic T.B., Glibetic M.D. Consumption of pomegranate juice decreases blood lipid peroxidation and levels of arachidonic acid in women with metabolic syndrome. J. Sci. Food Agric. 2017 97,1798-1804. doi: 10.1002/jsfa.7977.

[49] Shema-Didi L., Sela S., Ore L., Shapiro G., Geron R., Moshe G., Kristal B. (2012) One year of pomegranate juice intake decreases oxidative stress, inflammation, and incidence of infections in hemodialysis patients: A randomized placebo-controlled trial. Free Radic. Biol. Med. 53,297-304. doi: 10.1016/j.freeradbiomed.2012.05.013.

[50] Singleton VL, Rossi JA (1965) Colorimetry of total phenolics with phosphomolybdic-phosphotungstic acid reagents. Am J Enol Vitic 16,144-158

Chapter 5

Could Pomegranate Fight against SARS-CoV-2?

Sally Elnawasany

Abstract

Pomegranate, *Punica granatum* L., is an authentic, generous fruit which is cultivated in many parts of the world for thousand years. The divine fruit was born from nature to provide humanity with its effluent benefits for life and health. Through the ages, Pomegranate occupied an eminent place in ayurvedic medicine. It was prescribed for treatment of parasitic infection, diarrhea, and ulcers. Pomegranate wealth of prolific pharmacological activities makes it a rich culture for multiple studies in recent years. It will not be surprising if Pomegranate provides humans with a possible help in SARS-CoV-2 pandemic. The enemy that has raided the world since the end of 2019.

Keywords: ayurvedic medicine, phytochemicals, pomegranate, SARS-CoV-2

1. Introduction

Pomegranate (*Punica granatum* L.) is a common authentic fruit that is consumed for its health benefits in the globe. It contains many phytochemical constituents mainly Phenolic compounds which are responsible for most of its pharmacological properties [1, 2]. Several studies roamed in the Pomegranate field for its therapeutic benefits; anti- inflammatory, anti -oxidant, anti-cancer, anti- viral and immune modulation activities [3]. The fact that put pomegranate on the top of phytochemical agents with possible anti SARS-CoV-2 potential. Which has been attacking the earth for over a year. It is one of Coronaviruses, member of the subfamily Coronavirinae in the family Coronaviridae and the order Nidovirales [4].

2. Severe acute respiratory syndrome coronavirus-2, SARS-CoV-2

Corona viruses are wide group of viruses of humans as well as some animals. The clinical impact is ranged from mild to severe respiratory disease. In the last two decades, the world faced two aggressive coronaviruses: severe acute respiratory syndrome coronavirus (SARS-COV) in 2002 and Middle East respiratory syndrome coronavirus (MERS-COV) in 2012 [4]. At the end of 2019, SARS-CoV-2 was reported in China, as an abnormal highly contagious viral pneumonia. Then shortly, the virus invaded the whole world [5, 6]. SARS-CoV-2 is an enveloped positive-sense single stranded RNA virus. It consists of four subunits, spike (S) glycoprotein, small envelope (E) glycoprotein, membrane (M) glycoprotein and nucleocapsid (N) protein [7]. Spike S protein with its two subunits, S1 and S2 is responsible for epithelial cell entry after its attachment to Angiotensin Converting

Enzyme 2, ACE2 receptors which is widely present in the respiratory tract and other parts of the body [8, 9]. While surface S1 subunit (specifically at receptor-binding domain, RDB region) attach to ACE2 receptor, transmembrane subunit (S2) starts membrane fusion between the virus and epithelial cell and begins endocytosis. This process is enabled by the two host cell enzymes; furin and transmembrane serine protease 2 (TMPRSS2) that cleaves S glycoprotein at S1/S2 [10, 11]. Then SARS-CoV-2 replicates and spreads down to the airways and occupies alveolar epithelial cells. Viral replication induces Intense immune response (Cytokine storm syndrome) with subsequent acute respiratory distress syndrome and respiratory failure, the main cause of death [12]. Treating SARS-CoV-2 infection is not easy, as we have not only to fight the virus and manage its respiratory sequalae, but we need to downregulate the hyper stimulated immune response as well. For this war, many agents have been recruited in different ways. Starting from Inhibition of virus entry as Umifenovir (Arbidol) that interferes with interaction between the viral S protein and ACE2 and block membrane fusion [13, 14]. Chloroquine and hydroxy-chloroquine (two drugs of plant origin) are also thought to inhibit viral entry but with controversial results [15, 16]. Using soluble recombinant hACE2, specific monoclonal antibodies to occupy ACE2 receptors is another method to counter act viral entry [17, 18]. Inhibition of virus replication is another modality for treatment. There are numerous trials on remedesivir [15], favipiravir [19], ribavirin, lopinavir and ritonavir to inhibit viral replication [20]. Since SARS-CoV-2 over stimulates the immune response causing what is called, cytokine storm syndrome [21]. Immune modulation is a promising target for treatment. Dexamethasone decreased mortality in mechanically ventilated and oxygen receiving patients [22]. Plasma from recovered patients, convalescent plasma-derived hyperimmune globulin and monoclonal antibodies targeting SARS-CoV-2 were also tried in many trials [23–26]. Interleukin-6 (IL-6) has an important role in the inflammatory response. Tocilizumab, interleukin-6 (IL-6) receptor-specific antibody downregulated the immune response in small trials [27, 28]. Moreover, inhibition of pro inflammatory Complement 5 by Eculizumab, a specific monoclonal antibody, helped to decrease pulmonary oedema in severe COVID-19 patients [29]. Interferon plays a role in reducing of viral replication, type I interferons provide a treatment options in COVID-19 infection [30, 31]. Protein kinases inhibitors as Baricitinib, a reversible Janus-associated kinase (JAK)-inhibitor can help in SARS-CoV-2 treatment through its anti-inflammatory, anti-viral and antifibrotic properties [32]. Baricitinib attenu-ated cytokine signaling in COVID-19 immune response. It also interfered with viral cell entry [33]. In another study, it improved with corticosteroids the respiration in SARS-CoV-2 pneumonia [34]. In addition, the Abl tyrosine kinase inhibitor (ATKI), imatinib was found to block viral fusion through attachment to receptor-binding domain (RBD) of SARS-CoV-2 spike protein [35]. In spite of all the previous treatment modalities, there is no proven curative agent for SARS-CoV-2 infection [36]. Which necessitates a continuous and hard search for new therapeutic agents including natural agents.

3. Could pomegranate fight against SARS-CoV-2?

3.1 Anti-viral action of pomegranate

Pomegranate attenuates many viruses [37]. Polyphenols and ellagic acid were proved to neutralized envelope virus via binding to the envelope lipid or sugar moieties [38]. Pomegranate juice succeeded to prevent Human immune deficiency virus-1 (HIV-1) cell entry by blocking CD4 and coreceptors CXCR4/CCR5 binding [39]. Ellagitannins of Pomegranate extract; punicalagin, punicalin and ellagic acid blocked the HCV NS3/4A protease activity in an in vitro study [40]. Furthermore, the activity of adenovirus was suppressed by Pomegranate peel ethanol extract on HeLa cell line. The 50% inhibitory concentration (IC50) and 50% Cytotoxicity Concentration (CC50) were 165 ± 10.1 and 18.6 ± 6.7 μg/ml, respectively [41]. In addition, Pomegranate juice and pomegranate polyphenol extract reduced viral titer of noroviruses with other foodborne viral surrogates [42]. A viricidal effect of Pomegranate powder extract with 800 μg/ml polyphenols was clarified. When the titer of influenza virus (PR8 (H1N1), X31 (H3N2), and a reassortant H5N1 virus of human isolate lowered by 3log in 5 min treatment at room temperature. This effect was explored by electron microscopy when disruption of viral structure appeared [43]. Moreover, the replication and agglutination of chicken RBC's by influenza virus was inhibited by Punicalagin, a phenol in pomegranate extract. Synergistic effect was noticed in oseltamivir combination [44]. The anti Influenza mechanism was emphasized in another study where Pomegranate peel ethyl alcohol extract (PPE) inhibited the influenza virus adsorption and replication though attenuation of viral polymerase activity and protein expression [45].

3.2 Immune modulatory action of pomegranate

Mast cells and basophils have a crucial role in inflammatory and immune response [46]. These cells release pro-inflammatory cytokines TNF-α, IL-6, IL-8, histamine which initiate acute- and late-phase inflammatory response [47]. Cytokine expression is induced by many pathways such as extra-cellular signal-regulated kinase (ERK), and c-Jun N-terminal kinase (JNK) and Nuclear factor (NF)-κB [48–50]. Immune modulation action of pomegranate was confirmed in many studies. Pomegranate fruit extract strongly attenuated phorbol-12-myristate 13-acetate plus calcium inophore A23187 (PMACI) induced inflammatory gene expression and reduced the release of interleukin (IL)-6 and IL-8 in the myeloid pre-cursor cell line KU812 cells. Through its action on c-jun N-terminal kinase (JNK), extracellular-regulated kinase (ERK) and Neucular factor Kappa β (NF-κB) dependent pathways [51]. NF-κB signaling stimulation is mediated by IL-1β binding to its specific cell surface receptor that activates IKKs with subsequent phosphorylation and degradation of IκB. This cascade was suppressed in human chondrocyte by pomegranate extract. Which interfered with the mRNA and protein expression of IL-6 and downregulated the activation of NF-κB/p65. Through inhibition of the IL-1β-mediated phosphorylation of IKKβ, expression of IKKβ mRNA and degradation of IκBα [52]. Moreover, in another in vitro study, Pomegranate flower (PFE) ethanol extract reduced IL-6, IL-1β and TNF-α production with IC_{50} value of 48.7, 71.3 and 62.5 μg/mL respectively, in lipo-poly saccharides (LPS) -induced RAW264.7 cell macrophage. This effect was attributed to inhibition of phosphorylation of mitogen-activated protein kinase (MAPK) subgroups, extracellular signal-regulated kinase (ERK), c-Jun N-terminal kinase (JNK) and P38 and translocation of the NF-B p65 subunit [53]. Pomegranate peel extract decreased the secretion of CXCL8 in both Caco-2 cells and colonic explants. Furthermore, it attenuated the expression of IL 1A, IL 6 and CXCL8 in lipopoly saccharide, LPS stimulated colonic tissues at a concentration of 5 g/ml [54].

3.3 Anti-tyrosine kinase action of pomegranate

Janus kinase (JAK) is a member of the non-receptor tyrosine kinase family. It triggers many inflammatory signaling pathways like signal transducer and activation of transcription (STAT) that induce chemotaxis of inflammatory cells such as mast cells, T cell, B cells, macrophages [55]. Baricitinib is a Janus kinase (JAK) inhibitor and is a numb-associated kinase NAK inhibitor which attenuates AP2-associated protein kinase-1 (AAK1), the protein that promotes viral endocytosis [56, 57]. Fortunately, Pomegranate shares Baricitinib its janus kinase inhibitory action. The fact that introduces Pomegranate as a possible treating agent of SARS-CoV-2. This action was highlighted in a study where Pomegranate leaf extract antagonized Janus Kinase1 (JAK1) enzyme activity in macrophage raw cells [58]. In another study, among ellagitannins containing fruits, pomegranate was the superior in JAK2 inhibition [59].

3.4 Anti-converting enzyme (ACE) action of Pomegranate

The renin-angiotensin-aldosterone system, RAAS organizes blood pressure, fluid balance and controls the vascular response to inflammation [60]. Imbalance in that system induces hypertension, fluid retention, and inflammatory and thrombotic complications [61]. Juxtaglomerular apparatus of the kidney secretes renin which acts on angiotensinogen to form Angiotensin I (A1). Angiotensin-converting enzyme (ACE) breaks AI to AII. Angiotensin II is the main controlling agent of RAAS through stimulation of type 1 receptor (AT_1 receptor) with subsequent vasoconstriction, water retention and inflammation. While The type 2 receptor, ATR2 counteract these effects [62]. ACE2 counterbalance ACE actions. It breaks down AI into angiotensin 1-9(A1-9), and AII into angiotensin 1-7(A1-7) which has vasodilator and anti-proliferative action [63]. The renin-angiotensin system is claimed to induce severe acute lung injury in SARS-CoV-2 infection and ACE2 protects against acute lung failure and its deficiency is associated with lung damage [64]. Binding of SARS-Cov-2 to ACE2 receptor attenuates ACE2 action with subsequent lung damage [65]. On that basis, soluble ACE2 was supposed to be a possible approach for coronavirus infection [66]. It was speculated that, The use of ACE Inhibitors is associated with increased concentration of angiotensin I which upregulates ACE2 [67]. It is ambiguous, whether this postulate increases the probability of SARS-CoV-2 infection. Or ACE2 upregulation will be beneficial for counterbalance the ACE2 virus-induced downregulation with improvement of lung defense [65]. Although the role of Angiotensin converting enzyme, ACE inhibitors in SARS-CoV-2 infection is controversial, Pomegranate has the potential of ACE inhibition and may help in this battle. *Punica granatum* juice extract lowered ACE level and mean arterial blood pressure. When it was given in a dose of (PJ- 100 mg/kg and 300 mg/kg: p.o.) in angiotensin-II treated rats for 4 weeks [68]. Similar effect was emphasized in a parallel study. When Pomegranate peel extract was administered to female rats for 30 days. It inhibited coronary angiotensin-converting enzyme (ACE) activity and oxidative stress [69]. In a clinical trial, Pomegranate juice reduced serum ACE activity and systolic blood pressure in hypertensive patients when it was consumed for 2 weeks at a dose of (50 ml, 1.5 mmol of total polyphenols per day) [70].

3.5 Anti-SARS-CoV-2 action of pomegranate

Pomegranate potentials against SARS-CoV-2 infection have been investigated in many studies.

3.5.1 Anti-SARS-CoV-2 action of pomegranate

In a Computational study, Pomegranate peel extracts components; ellagic acid, gallic acid and specially punicalagin, punicalin showed promising anti SARS-CoV-2 activity through interaction with SARS-CoV-2 spike glycoprotein, angiotensin converting enzyme 2, furin and transmembrane serine protease2. They formed more stable complexes with amino acid residues at the active sites of the selected protein targets in comparison to positive controls (umifenovir, lopinavir, camostat) with more significant binding affinity. Punicalin showed the most potent interaction with the S glycoprotein with free binding energy of −7.406 kcal/mol. All Pomegranate components ligands exerted a significant binding affinity at the ACE2 predicted active site. Furthermore, they formed the most stable complexes with furin. Amazingly, Punicalagin and punicalin strongly interacted with TMPRSS2 amino acid residues at the predicted active site by binding energy values of −7.358 and − 8.168 kcal/mol, respectively with higher affinity for the target protein than camostat (−7.069 kcal/mol) [71]. In an in vitro study, Pomegranate juice reduced the infectivity of SARS-Cov-2 and influenza virus in VeroE6 cells [72]. In another study, Pomegranate peel extract showed an ability to block the binding between SARS-CoV-2 Spike glycoprotein and the human Angiotensin-Converting Enzyme 2 (ACE2) receptor, furthermore, it downregulated the activity of the virus bind 3-chymotrypsin-like cysteine protease ($3CL^{Pro}$) (an enzyme which is important for viral replication) [73].

3.5.2 Anti-SARS-CoV-2 action of natural compounds that are found in pomegranate

In a virtual study, pedunculagin, tercatain, and castalin (hydrolysable tannins) showed an ability to bind ($3CL^{Pro}$) catalytic site that is involved in SARS-CoV-2 replication. Which sheds the light on tannins as possible anti SARS-CoV-2 agents [74]. Other virtual study investigated the action of natural compounds on SARS-CoV-2 Spike protein, viral Protease and RNA-dependent RNA polymerase and host cell protease TMPRSS2. Triterpenoids was found to be the superior in blocking the Spike protein binding site of SARS-CoV-2 [75].

4. Conclusion

Pomegranate is still surprising the world by its great therapeutic benefits. This chapter highlights the anti-SARS-CoV-2 potentials of Pomegranate. Where Antiviral, immune modulation, tyrosine kinase and ACE inhibition actions, all enable Pomegranate to fight in this war. Further future studies are needed to confirm the utility of Pomegranate in treating SARS-CoV-2 infection.

Conflict of interest

I confirm that there are no conflicts of interest.

Author details

Sally Elnawasany
Tropical Medicine, Tanta University, Egypt

*Address all correspondence to: elnawasany_s@hotmail.com

IntechOpen

References

[1] Ismail T, Akhtar S, Riaz M.
Pomegranate Peel and Fruit Extracts: A
Novel Approach to Avert Degenerative
Disorders–Pomegranate and
Degenerative Diseases. InExploring
the Nutrition and Health Benefits of
Functional Foods 2017 (pp. 165-184).
IGI Global.

[2] Goertz A, Ahmad KA. Biological
activity of phytochemical compounds in
pomegranate-a review. EC Nutri-tion.
2015;1:115-127.

[3] Elnawasany S. Clinical Applications
of Pomegranate. In: Soneji J,
Nageswara-Rao M, editors. Breeding
and Health Benefits of Fruit and Nut
Crops, Intechopen; 2018. p. 127-148.
DOI: 10.5772/intechopen. 75962

[4] Cui J, Li F, Shi ZL. Origin and
evolution of pathogenic coronaviruses.
Nature Reviews Microbiology. 2019
Mar;17(3):181-192.

[5] Wu JT, Leung K, Leung GM.
Nowcasting and forecasting the
potential domestic and international
spread of the 2019-nCoV outbreak
originating in Wuhan, China: a
modelling study. The Lancet. 2020 Feb
29;395(10225):689-697.

[6] Hui DS, Azhar EI, Madani TA,
Ntoumi F, Kock R, Dar O, Ippolito G,
Mchugh TD, Memish ZA, Drosten C,
Zumla A. The continuing 2019-nCoV
epidemic threat of novel coronaviruses
to global health—The latest 2019 novel
coronavirus outbreak in Wuhan, China.
International Journal of Infectious
Diseases. 2020 Feb 1;91:264-266.

[7] Jiang S, Hillyer C, Du L. Neutralizing
antibodies against SARS-CoV-2 and
other human coronaviruses. Trends in
immunology. 2020 Apr 2

[8] Rabi FA, Al Zoubi MS,
Kasasbeh GA, Salameh DM,

Al-Nasser AD. SARS-CoV-2 and
coronavirus disease 2019: what we know
so far. Pathogens. 2020 Mar;9(3):231.

[9] Boström KI, Yao Y. Options for
COVID-19 Entry into Pulmonary
Cells. Biomedical journal of
scientific & technical research. 2020
Aug;29(2):22337.

[10] Hoffmann M, Hofmann-Winkler H,
Pöhlmann S. Priming time: How
cellular proteases arm coronavirus
spike proteins. InActivation of Viruses
by Host Proteases 2018 (pp. 71-98).
Springer, Cham.a

[11] Shang J, Ye G, Shi K, Wan Y, Luo C,
Aihara H, Geng Q, Auerbach A, Li F.
Structural basis of receptor recognition
by SARS-CoV-2. Nature. 2020
May;581(7807):221-224.

[12] Tufan A, GÜLER AA,
Matucci-Cerinic M. COVID-19, immune
system response, hyperinflammation
and repurposing antirheumatic drugs.
Turkish Journal of Medical Sciences.
2020 Apr 21;50(SI-1):620-632.

[13] Wang X, Cao R, Zhang H, Liu J,
Xu M, Hu H, Li Y, Zhao L, Li W, Sun X,
Yang X. The anti-influenza virus drug,
arbidol is an efficient inhibitor of SARS-
CoV-2 in vitro. Cell Discovery. 2020
May 2;6(1):1-5.

[14] Zhu Z, Lu Z, Xu T, Chen C, Yang G,
Zha T, Lu J, Xue Y. Arbidol monotherapy
is superior to lopinavir/ritonavir in
treating COVID-19. Journal of Infection.
2020 Jul 1;81(1):e21-e23.

[15] Wang M, Cao R, Zhang L, Yang X,
Liu J, Xu M, Shi Z, Hu Z, Zhong W,
Xiao G. Remdesivir and chloroquine
effectively inhibit the recently emerged
novel coronavirus (2019-nCoV) in vitro.
Cell research. 2020 Mar;30(3):269-271.

[16] Yao X, Ye F, Zhang M, Cui C,
Huang B, Niu P, Liu X, Zhao L,

Dong E, Song C, Zhan S. In vitro antiviral activity and projection of optimized dosing design of hydroxychloroquine for the treatment of severe acute respiratory syndrome coronavirus 2 (SARS-CoV-2). Clinical Infectious Diseases. 2020 Mar 9.

[17] Monteil V, Kwon H, Prado P, Hagelkrüys A, Wimmer RA, Stahl M, Leopoldi A, Garreta E, Del Pozo CH, Prosper F, Romero JP. Inhibition of SARS-CoV-2 infections in engineered human tissues using clinical-grade soluble human ACE2. Cell. 2020 Apr 24.

[18] Tian X, Li C, Huang A, Xia S, Lu S, Shi Z, Lu L, Jiang S, Yang Z, Wu Y, Ying T. Potent binding of 2019 novel coronavirus spike protein by a SARS coronavirus-specific human monoclonal antibody. Emerging microbes & infections. 2020 Jan 1;9(1):382-385.

[19] Agrawal U, Raju R, Udwadia ZF. Favipiravir: A new and emerging antiviral option in COVID-19. Medical Journal Armed Forces India. 2020 Sep 2.

[20] Zeng YM, Xu XL, He XQ, Tang SQ, Li Y, Huang YQ, Harypursat V, Chen YK. Comparative effectiveness and safety of ribavirin plus interferon-alpha, lopinavir/ritonavir plus interferon-alpha, and ribavirin plus lopinavir/ ritonavir plus interferon-alpha in patients with mild to moderate novel coronavirus disease 2019: study protocol. Chinese medical journal. 2020 May 5;133(9):1132-1134.

[21] Mehta P, McAuley DF, Brown M, Sanchez E, Tattersall RS, Manson JJ, HLH Across Speciality Collaboration. COVID-19: consider cytokine storm syndromes and immunosuppression. Lancet (London, England). 2020 Mar 28;395(10229):1033.

[22] Ahmed MH, Hassan A. Dexamethasone for the Treatment of Coronavirus Disease (COVID-19): a

Review. SN comprehensive clinical medicine. 2020 Oct 31:1-0.

[23] Shen C, Wang Z, Zhao F, Yang Y, Li J, Yuan J, Wang F, Li D, Yang M, Xing L, Wei J. Treatment of 5 critically ill patients with COVID-19 with convalescent plasma. Jama. 2020 Apr 28;323(16):1582-1589.

[24] Li L, Zhang W, Hu Y, Tong X, Zheng S, Yang J, Kong Y, Ren L, Wei Q, Mei H, Hu C. Effect of Convalescent Plasma Therapy on Time to Clinical Improvement in Patients With Severe and Life-threatening COVID-19: A Randomized Clinical Trial. Jama. 2020 Jun 3.

[25] Wang C, Li W, Drabek D, Okba NM, van Haperen R, Osterhaus AD, van Kuppeveld FJ, Haagmans BL, Grosveld F, Bosch BJ. A human monoclonal antibody blocking SARS-CoV-2 infection. Nature communications. 2020 May 4;11(1):1-6.

[26] Brouwer P, Caniels T, van Straten K, Snitselaar J, Aldon Y, Bangaru S, Torres J, Okba N, Claireaux M, Kerster G, Bentlage A. Potent neutralizing antibodies from COVID-19 patients define multiple targets of vulnerability. bioRxiv. 2020 Jan 1.

[27] Xu X, Han M, Li T, Sun W, Wang D, Fu B, Zhou Y, Zheng X, Yang Y, Li X, Zhang X. Effective treatment of severe COVID-19 patients with tocilizumab. Proceedings of the National Academy of Sciences. 2020 May 19;117(20):10970-10975.

[28] Alzghari SK, Acuña VS. Supportive treatment with tocilizumab for COVID-19: a systematic review. Journal of Clinical Virology. 2020 Jun;127:104380.

[29] Diurno F, Numis FG, Porta G, Cirillo F, Maddaluno S, Ragozzino A, De Negri P, Di Gennaro C, Pagano A, Allegorico E, Bressy L. Eculizumab

treatment in patients with COVID-19: preliminary results from real life ASL Napoli 2 Nord experience. European Review for Medical and Pharmacological Sciences. 2020 Apr 1;24(7):4040-4047.

[30] Stockman LJ, Bellamy R, Garner P. SARS: systematic review of treatment effects. PLoS Med. 2006 Sep 12;3(9):e343.

[31] Mantlo E, Bukreyeva N, Maruyama J, Paessler S, Huang C. Antiviral activities of type I interferons to SARS-CoV-2 infection. Antiviral research. 2020 Apr 29:104811.

[32] Weisberg E, Parent A, Yang PL, Sattler M, Liu Q, Liu Q, Wang J, Meng C, Buhrlage SJ, Gray N, Griffin JD. Repurposing of kinase inhibitors for treatment of COVID-19. Pharmaceutical research. 2020 Sep;37(9):1-29.

[33] Jorgensen SC, Tse CL, Burry L, Dresser LD. Baricitinib: a review of pharmacology, safety, and emerging clinical experience in COVID-19. Pharmacotherapy: The Journal of Human Pharmacology and Drug Therapy. 2020 Aug;40(8):843-856.

[34] Rodriguez-Garcia JL, Sanchez-Nievas G, Arevalo-Serrano J, Garcia-Gomez C, Jimenez-Vizuete JM, Martinez-Alfaro E. Baricitinib improves respiratory function in patients treated with corticosteroids for SARS-CoV-2 pneumonia: an observational cohort study. Rheumatology. 2021 Jan;60(1):399-407.

[35] Mulgaonkar NS, Wang H, Mallawarachchi S, Ruzek D, Martina B, Fernando S. Bcr-Abl tyrosine kinase inhibitor imatinib as a potential drug for COVID-19. bioRxiv. 2020 Jan 1.

[36] Janković S. Current status and future perspective of coronavirus disease 2019: a review. Scripta Medica. 2020;51(2):101-109.

[37] Howell AB, D'Souza DH. The pomegranate: effects on bacteria and viruses that influence human health. Evidence-Based Complementary and Alternative Medicine. 2013 Oct;2013.

[38] Kotwal GJ. Genetic diversity-independent neutralization of pandemic viruses (eg HIV), potentially pandemic (eg H5N1 strain of influenza) and carcinogenic (eg HBV and HCV) viruses and possible agents of bioterrorism (variola) by enveloped virus neutralizing compounds (EVNCs). Vaccine. 2008;26(24):3055-3058

[39] Neurath AR, STRICK N, LI YY, DEBNATH AK. Punica granatum (pomegranate) juice provides an HIV-1 entry inhibitor and candidate topical microbicide. Annals of the New York Academy of Sciences. 2005 Nov;1056(1):311-27.

[40] Reddy BU, Mullick R, Kumar A, Sudha G, Srinivasan N, Das S. Small molecule inhibitors of HCV replication from pomegranate. Scientific reports. 2014 Jun 24;4:5411.

[41] Karimi A, Moradi MT, Rabiei M, Alidadi S. In vitro anti-adenoviral activities of ethanol extract, fractions, and main phenolic compounds of pomegranate (Punica granatum L.) peel. Antiviral Chemistry and Chemotherapy. 2020 Apr;28:2040206620916571.

[42] Su X, Sangster MY, D'Souza DH. In vitro effects of pomegranate juice and pomegranate polyphenols on foodborne viral surrogates. Foodborne Pathogens and Disease. 2010 Dec 1;7(12):1473-1479.

[43] Sundararajan A, Ganapathy R, Huan L, Dunlap JR, Webby RJ, Kotwal GJ, Sangster MY. Influenza virus variation in susceptibility to inactivation by pomegranate polyphenols is determined by envelope glycoproteins. Antiviral research. 2010 Oct 1;88(1):1-9.

[44] Haidari M, Ali M, Casscells III SW, Madjid M. Pomegranate (Punica granatum) purified polyphenol extract inhibits influenza virus and has a synergistic effect with oseltamivir. Phytomedicine. 2009 Dec 1;16(12):1127-1136.

[45] Moradi MT, Karimi A, Rafieian-kopaei M, Rabiei-Faradonbeh M, Momtaz H. Pomegranate peel extract inhibits internalization and replication of the influenza virus: An in vitro study. Avicenna Journal of Phytomedicine. 2020 Mar;10(2):143.

[46] Galli SJ. New concepts about the mast cell. New England Journal of Medicine. 1993 Jan 28;328(4):257-265.

[47] Woolley DE, Tetlow LC. Mast cell activation and its relation to proinflammatory cytokine production in the rheumatoid lesion. Arthritis Research & Therapy. 1999 Dec;2(1):1-0.

[48] Cobb MH, Goldsmith EJ. Dimerization in MAP-kinase signaling. Trends in biochemical sciences. 2000 Jan 1;25(1):7-9.

[49] Lewis TS, Shapiro PS, Ahn NG. Signal transduction through MAP kinase cascades. In Advances in cancer research 1998 Jan 1 (Vol. 74, pp. 49-139). Academic Press.

[50] Collart MA, Baeuerle P, Vassalli P. Regulation of tumor necrosis factor alpha transcription in macrophages: involvement of four kappa B-like motifs and of constitutive and inducible forms of NF-kappa B. Molecular and cellular biology. 1990 Apr 1;10(4):1498-1506.

[51] Rasheed Z, Akhtar N, Anbazhagan AN, Ramamurthy S, Shukla M, Haqqi TM. Polyphenol-rich pomegranate fruit extract (POMx) suppresses PMACI-induced expression of pro-inflammatory cytokines by inhibiting the activation of MAP Kinases and NF-κB in human KU812

cells. Journal of inflammation. 2009 Dec 1;6(1):1.

[52] Haseeb A, Khan NM, Ashruf OS, Haqqi TM. A polyphenol-rich pomegranate fruit extract suppresses NF-κB and IL-6 expression by blocking the activation of IKKβ and NIK in primary human chondrocytes. Phytotherapy Research. 2017 May;31(5):778-782.

[53] Xu J, Zhao Y, Aisa HA. Anti-inflammatory effect of pomegranate flower in lipopolysaccharide (LPS)-stimulated RAW264. 7 macrophages. Pharmaceutical Biology. 2017 Jan 1;55(1):2095-2101.

[54] Mastrogiovanni F, Mukhopadhya A, Lacetera N, Ryan MT, Romani A, Bernini R, Sweeney T. Anti-inflammatory effects of pomegranate peel extracts on in vitro human intestinal caco-2 cells and ex vivo porcine colonic tissue explants. Nutrients. 2019 Mar;11(3):548.

[55] Wong WF, Leong KP. Tyrosine kinase inhibitors: a new approach for asthma. Biochimica et Biophysica Acta (BBA)-Proteins and Proteomics. 2004 Mar 11;1697(1-2):53-69.

[56] Stebbing J, Phelan A, Griffin I, Tucker C, Oechsle O, Smith D, Richardson P. COVID-19: combining antiviral and anti-inflammatory treatments. The Lancet Infectious Diseases. 2020 Apr 1;20(4):400-402.

[57] Richardson P, Griffin I, Tucker C, Smith D, Oechsle O, Phelan A, Stebbing J. Baricitinib as potential treatment for 2019-nCoV acute respiratory disease. Lancet (London, England). 2020 Feb 15;395(10223):e30.

[58] Sarithamol S, Pushpa VL, Manoj KB. Comparative Study on Janus Kinase Enzyme activity of Pomegranate Leaf Extract and its Active Component

Ellagic Acid for Asthma. Oriental Journal of Chemistry. 2018;34(2):1041.

[59] Martin H, Burgess EJ, Smith WA, McGhie TK, Cooney JM, Lunken RC, de Guzman E, Trower T, Perry NB. JAK2 and AMP-kinase inhibition in vitro by food extracts, fractions and purified phytochemicals. Food & function. 2015;6(1):304-311.

[60] Ferrario CM. Role of angiotensin II in cardiovascular disease—therapeutic implications of more than a century of research. Journal of the Renin-angiotensin-aldosterone System. 2006 Mar;7(1):3-14.

[61] Brewster UC, Perazella MA. The renin-angiotensin-aldosterone system and the kidney: effects on kidney disease. The American journal of medicine. 2004 Feb 15;116(4):263-272.

[62] Zaman MA, Oparil S, Calhoun DA. Drugs targeting the renin–angiotensin–aldosterone system. Nature reviews Drug discovery. 2002 Aug;1(8):621-636.

[63] Macia-Heras M, Del Castillo-Rodriguez N, Navarro González J. The renin-angiotensin-aldosterone system in renal and cardiovascular disease and the effects of its pharmacological blockade. J Diabetes Metab. 2012 Feb 1;3(2).

[64] Kuba K, Imai Y, Rao S, Gao H, Guo F, Guan B, Huan Y, Yang P, Zhang Y, Deng W, Bao L. A crucial role of angiotensin converting enzyme 2 (ACE2) in SARS coronavirus–induced lung injury. Nature medicine. 2005 Aug;11(8):875-879.

[65] Perrotta F, Matera MG, Cazzola M, Bianco A. Severe respiratory SARS-CoV2 infection: Does ACE2 receptor matter?. Respiratory Medicine. 2020 Apr 25:105996.

[66] Batlle D, Wysocki J, Satchell K. Soluble angiotensin-converting enzyme 2: a potential approach for coronavirus infection therapy?. Clinical science. 2020 Mar 13;134(5):543-545.

[67] Tikellis C, Thomas MC. Angiotensin-converting enzyme 2 (ACE2) is a key modulator of the renin angiotensin system in health and disease. International journal of peptides. 2012;2012.

[68] Harshal W, Mahalaxmi M, Sanjay K, Balaraman R. Punica granatum attenuates angiotensin-II induced hypertension in Wistar rats. International Journal of PharmTech Research. 2010;2(1):60-67.

[69] dos Santos RL, Dellacqua LO, Delgado NT, Rouver WN, Podratz PL, Lima LC, Piccin MP, Meyrelles SS, Mauad H, Graceli JB, Moyses MR. Pomegranate peel extract attenuates oxidative stress by decreasing coronary angiotensin-converting enzyme (ACE) activity in hypertensive female rats. Journal of Toxicology and Environmental Health, Part A. 2016 Nov 1;79(21):998-1007.

[70] Aviram M, Dornfeld L. Pomegranate juice consumption inhibits serum angiotensin converting enzyme activity and reduces systolic blood pressure. Atherosclerosis. 2001 Sep 1;158(1):195-198.

[71] Suručić R, Tubić B, Stojiljković MP, Djuric DM, Travar M, Grabež M, Šavikin K, Škrbić R. Computational study of pomegranate peel extract polyphenols as potential inhibitors of SARS-CoV-2 virus internalization. Molecular and cellular biochemistry. 2020 Nov 16:1-5.

[72] Conzelmann C, Weil T, Groß R, Jungke P, Frank B, Eggers M, Müller JA, Münch J. Antiviral activity of plant juices and green tea against SARS-CoV-2 and influenza virus in vitro. bioRxiv. 2020 Jan 1.

[73] Tito A, Colantuono A, Pirone L, Pedone EM, Intartaglia D, Giamundo G, Conte I, Vitaglione P, Apone F. A pomegranate peel extract as inhibitor of SARS-CoV-2 Spike binding to human ACE2 (in vitro): a promising source of novel antiviral drugs. bioRxiv. 2020 Jan 1.

[74] Khalifa I, Zhu W, Mohammed HH, Dutta K, Li C. Tannins inhibit SARS-CoV-2 through binding with catalytic dyad residues of 3CLpro: An in silico approach with 19 structural different hydrolysable tannins. Journal of food biochemistry. 2020 Oct;44(10):e13432.

[75] Gowtham HG, Monu DO, Ajay Y, Gourav C, Vasantharaja R, Bhani K, Koushalya S, Shazia S, Priyanka G, Leena C. Exploring structurally diverse plant secondary metabolites as a potential source of drug targeting different molecular mechanisms of Severe Acute Respiratory Syndrome Coronavirus-2 (SARS-CoV-2) pathogenesis: An in silico approach.

Section 3

Post-Harvest Technology of Pomegranate

Chapter 6

Post-Harvest Management and Value Addition in Pomegranate

Sangram S. Dhumal, Ravindra D. Pawar and Sandip S. Patil

Abstract

Pomegranate due to its high nutritive and therapeutic value, high antioxidant capacity, and bioactive compounds is known as superfruit. However, its consumption is scarce due to difficulties in peeling and extraction of arils, hand staining and irritation during extraction due to phenolic metabolites in fruits. Improved varieties have excellent flavour with crisp-juicy-dark red, gem-like arils, indicating potentiality for export and value-added products with the extended shelf life. Advances in post-harvest technology had played a vital role in product diversification by keeping original nutritional value. Extensive research has been carried out in the development of various pomegranate-derived products such as minimally processed arils, frozen seeds, RTS juice, concentrates. These processed products are highly acceptable because of their dessert qualities and palatability. Consumers readily pick well-matured big size fruits with attractive colour but low-grade pomegranate is kept out of market. Additional innovative tools like modified atmosphere packaging offer for the optimal use of such lower-grade fruits. Consumers prefer minimally processed pomegranate arils and frozen arils packed in punnets over whole fruit. Juices can be used in beverages and for various treatment purposes. This new sector of pomegranate processing will allow the use of non-commercial pomegranate fruits and improve pomegranate utilization for human health.

Keywords: grading, packing, storage, cool chain, processing, value addition, molasses, juice, arils RTS

1. Introduction

Pomegranate has been cultivated in India, South East Asia and tropical continents for thousands of years. Previously pomegranate was not grown as table fruit, but it was grown as decorative plant with its ornamental value, for its red, orange or, occasionally, creamy yellow flowers. Pomegranate is also mentioned in biblical writings and was used as blessings in ceremonies. The pomegranate (*Punica granatum* L.) belongs to the Punicaceae family. It is a native of Iran to Himalayan region. Pomegranate is also known as the *Chinese apple or Apple of Carthage or Apple with many seeds*. It is grown in tropical and subtropical parts of the world and is extensively grown in Iran, Spain, India and USA as well as in most Near and Far East countries [1]. It is highly adaptive to the various climates and found to be commercially cultivated in diverse geographical regions including Mediterranean region, Asia and USA. Among all the pomegranate-producing countries across the world, the 76 per cent of the production is contributed by Iran, India, Spain, China, Turkey and USA.

As pomegranate thrives best under hot dry summer and cold winter provided irrigation facilities and can withstand different soil and climatic stresses, it is grown in the arid and semi-arid regions of the country [2]. Pomegranate has high nutritional values and it requires low maintenance cost, has high yielding potential and is tolerant to biotic and abiotic stresses making it popular among growers as well as consumers. It is grown in the foothills of Himalayas along entire hilly tract of Jammu and Kashmir, Himachal Pradesh and parts of Uttar Pradesh. India ranks first in the pomegranate production (28.45 lakh tonnes) in the world, on an area of 2.34 lakh hectares with productivity of 12.16 t/ha [3]. Maharashtra, Karnataka, Gujarat, Rajasthan, Uttar Pradesh, Andhra Pradesh and Tamil Nadu are the major pomegranate growing states in India. Maharashtra, which emerged as a pomegranate basket of India, has 1.47 lakh ha area (63 per cent) under pomegranate cultivation with 62.88 per cent (17.89 lakh tones) of share in total production of the country. Though the country is the largest producer of pomegranate, more than 90 per cent of the produce is utilized for domestic fresh consumption and remaining is exported. India's share in global exports is around 6.4 per cent as compared to Spain and Iran having global market share of 45 per cent and 15 per cent, respectively, competing India in international market. UAE, Oman, Saudi Arabia, the Netherlands, Qatar, Iran, Kuwait as well as to nearby countries like Nepal and Bangladesh are the major importers of pomegranate.

The fruit is valued for its high remunerative returns under a wide range of climatically challenging cultivation conditions. Pomegranate can be grown with less water, tolerates high-temperature variations and responds to high-tech horticultural practices. Excellent flavour, nutritive value and medicinal properties of pomegranate fruit indicate its good potentiality for processing into value-added products having extended shelf life. It can be processed into variety of products besides having fairly good export potential. Lack of technological developments for commercialization, resource personnel and scientific research database are the major reasons for the underdevelopment of pomegranate processing industry in India, in spite of the known nutraceutical benefits and great global demand for potentially pomegranate-derived products resulting in very high post-harvest losses (20–40 per cent). Improper post-harvest management adds to another 10–15 per cent fresh produce losing its market value and consumer acceptability. Minimizing these losses can increase pomegranate supply without bringing additional land under cultivation. In order to exploit and popularize the medicinal and nutritive values of the pomegranate to its fullest extent, it becomes essential to explore the different ways of minimal processing and post-harvest technology applications.

2. Therapeutic properties of pomegranate

Pomegranate is a very promising and emerging crop for its refreshing arils, juice and other products. It is well documented for its chemopreventive properties having medicinal value [4]. The pomegranate has been regarded as a food medicine of great importance for therapeutic purposes like colic, colitis-diarrhoea, dysentery, leucorrhea, paralysis and headache [1, 5]. The wide applications in the traditional Asian medicines both in Ayurvedic and Unani systems are well known. The therapeutic properties are reported to be due to the presence of betulinic and ursolic acids and different alkaloids such as pseudopelletierine, pelletierine and some other basic compounds [6]. There has been a remarkable increase in the commercial farming of the pomegranates globally due to the potential health benefits such as its high antioxidant, antimutagenic, and antihypertensive activities, and

Plant part	Constituents
Pomegranate juice	Anthocyanins, glucose, ascorbic acid, ellagic acid, gallic acid, caffeic acid, catechin, minerals, amino acids, quercetin and rutin
Pomegranate seed oil	95% punicic acid, ellagic acid and sterols
Pomegranate pericarp (peel and rind)	Phenolic punicalagins, gallic acid, catechin, flavones, flavonones and anthocyanidins
Pomegranate leaves	Tannins, falvone glycosides, luteolin and apigenin
Pomegranate flower	Gallic acid, ursolic acid, and triterpenoids including maslinic and asiatic acid
Pomegranate roots and bark	Ellagitannins, punicalin and punicalagin, piperidine alkaloids

Source: Jurenka [13].

Table 1.
Pomegranate plant parts and its constituents.

the ability to reduce liver injury [7–10]. The pomegranate anthocyanins and poly-phenolic compounds are known for scavenging activities and are able to elevate the antioxidant capacity of the human body. Pomegranate fruit is also known for its anti-inflammatory and anti-atherosclerotic activities against osteoarthritis, prostate cancer, heart disease and HIV-I [11, 12]. The juice from the pomegranates is one of the nature's most powerful antioxidants. Pomegranate juice also increases the body's resistance against infections, acts as cooling beverage and improves the function of kidney, liver and heart. All the parts of the tree, the roots, the reddish brown bark, leaves, flowers, rinds and seeds are reported to have rich source of different chemical constituents (**Table 1**) and thus included in medicines since thousand years. The sweet varieties of pomegranate are considered a good laxative, while those which are intermediate between sweet and sour are regarded as valuable in stomach inflammations and heart pain. The pomegranates have recently been found to boost activity of an enzyme which protects the cardiovascular risks.

3. Pomegranate fruit quality characteristics

Botanically, pomegranate is a fleshy berry, a balausta, and is considered to be a non-climacteric fruit. It has multiple ovule chambers separated by membranous walls (septum) and a fleshy mesocarp. The chambers are filled with shiny red seeds encased in succulent and edible red pink pulp called arils. However, the edible portion of pomegranate is testa [14]. The arils develop from the outer epidermal cells of the seed and elongate to a very large extent in a radial direction [15]. The colour of arils varies from white to deep red depending on the variety. The fruits are irregular round in shape and vary from yellow, green or pink to bright deep red in colour depending on the variety and stage of ripening [16]. Pomegranate has received special attention in recent years due to its properties, viz. rich in sugars, organic acids, minerals, anthocyanins, flavo-noids, punicic acid, the sex steroid estrone and the phytoestrogen coumestrol.

The arils, edible portion of pomegranate, consist of around 80 per cent juice and 20 per cent seed [17]. The quality of the pomegranate fruit largely depends on its size, skin colour the absence of visual defects, such as sunburn, cracks, cuts, bruises and decay, and also on the presence of small and soft seeds as well as on aril colour, sugar and acid content [18]. The edible portion of pomegranate is an excellent dietary source as it contains a significant proportion of organic acids, soluble solids, polysaccharides, vitamins, fatty acids and mineral elements of nutritional significance (**Tables 2** and **3**) [20, 21]. It has a wide range of significance in human health, nutritional and livelihood security, which has been recognized resulting in increased fruit consumption [22].

Constituents	Edible fruits	
	Fresh fruits	**Dry weight basis**
Moisture (%)	78	19
Protein (%)	1.6	7.27
Total sugars (%)	14.6	66.36
Ascorbic acid (mg/100 g)	16.0	72.73
Ash (%)	0.7	3.18
Acidity (%)	0.58	2.64
Mineral (mg/100 g)		
Calcium	10	45
Phosphorous	70	318
Magnesium	44	200
Potassium	133	604
Sodium	0.90	4.09
Iron	1.79	8.14
Zinc	0.82	3.73
Manganese	0.77	3.50
Copper	0.34	1.55
Source: Chavan et al. [19].		

Table 2.
Chemical and mineral composition of pomegranate fruits.

However, poor post-harvest handling, storage recommendations, short shelf life and quality deterioration during transport, storage and marketing resulted in the limited consumption despite the increasing consumer awareness of the health benefits of pomegranate, whereas the occurrence of physiological disorders, viz. husk scald, splitting and chilling injury, reduces marketability and consumers acceptance.

4. Pomegranate maturity indices

For getting the quality pomegranate fruits, they are harvested at proper stage of maturity. Under Indian conditions in Maharashtra, the maturity of pomegranate fruit is judged by the following indices:

1. The fruit requires 130–150 days after fruit set for proper maturity.

2. In summer, the fruit colour changes from yellowish to dark red at maturity.

3. The fruit is harvested when it gives cracking sound when tapped.

4. The calyx at the distal end of the fruit gets closed at the time of maturity.

5. Properly matured fruit is easily scratched with fingernails.

Constituents	Unit	Value per 100 gram of edible portion
Proximates		
Water	g	80.97
Energy	Kcal	68
Protein	g	0.95
Total lipid	g	0.30
Ash	g	0.61
Carbohydrate, by difference	g	17.17
Total dietary fibre	g	0.6
Total sugars	g	16.57
Minerals		
Calcium	mg	3
Iron	mg	0.30
Magnesium	mg	3
Phosphorous	mg	8
Potassium	mg	259
Sodium	mg	3
Zinc	mg	0.12
Copper	mg	0.070
Selenium	mcg	0.6
Vitamins		
Vitamin C (total ascorbic acid)	mg	6.1
Thiamin	mg	0.030
Riboflavin	mg	0.030
Niacin	mg	0.300
Pantothenic acid	mg	0.596
Vitamin B_6	mg	0.105
Total folate	mcg	6
Vitamin A	IU	108
Vitamin E	mg	0.60
Vitamin K	mcg	4.6

Source: USDA National Nutrient Database for Standard Reference, Release 17 (2004).

Table 3.
Nutritional composition of the pomegranate.

5. Harvesting

The overripe fruit affects the shelf life and also the cost. The mature fruit is harvested carefully with minimum damage. The picking of fruit with stalk about 1 cm intact helps to increase its storage life. Detachment of the fruit with sharp knife is also recommended. The picking of the fruit is done early in the morning or late evening. After picking, the fruits are collected in the plastic crates and cleaned with the help of clothes in order to remove the dust with minimum damage.

6. Grading

Proper grading of the fruits is an important practice to maintain the quality and to fetch the optimum price in domestic as well as in international markets. The pomegranate fruits are graded on the basis of weight, size and external colour. During grading, the malformed, diseased, immature and cracked fruits are grouped into the following grades:

   i. *Supersize*: The fruits of this grade have good attractive bright red colour with weight more than 750 g per fruit and do not have any spots on the skin.

   ii. *King size*: The fruits free from spots having an attractive red colour with weight ranging between 500 and 750 g come under this grade.

   iii. *Queen size*: The fruits free from spots having a bright red colour with weight ranging between 400 and 500 g come under this grade.

   iv. *Prince size*: The fully ripe fruits weighing between 300 and 400 g with red colour come under this grade.

Besides these grades, the fruits are also graded into two more grades such as 12-A grade and 12-B grade. The fruits weighing between 250 and 300 g with some spots on the skin are graded as 12-A grade. The fruits of this grade are generally preferred in South and North Indian markets.

7. Packaging and storage

In pomegranate, the size of packaging changes according to the grade of the fruits. Corrugated fibreboard boxes (CFBs) are used for packaging as they have many advantages such as lightweight, providing less or no damage to the fruits, easy to handle and reduction in the freight charges. In a single box, 4–5 fruits of Supersize, 6 fruits of King size, 9 fruits of Queen size and 12 fruits of Prince size and 12-A and 12-B grades are generally packed. The boxes with white colour having five plies are generally used for the export purpose, whereas the boxes with red colour with three plies are used for domestic markets. The red colour boxes are cheaper than the white coloured ones. The sizes of boxes used for packaging fruits of Supersize, Queen size and Prince size and 12-A and 12-B grades are 13 x 9 x 4 inches, 15 x 11 x 4 inches and 14 x 10 x 4 inches, respectively. During packing the fruits, the cut pieces of the waste paper are generally used as cushioning material. Then the graded fruits are placed on cushioning material followed by an attractive red colour paper on the boxes. Storage temperatures of 0–1.7°C and RH of 85–90 per cent are recommended for storing Kandahar pomegranates for 11 weeks [23].

It is reported that pomegranate cv. Banlaung were packed in plastic baskets and sealed with polyethylene (PE) bags (0.02 mm) and held at 5, 10 and 30°C (RT). The storage life of pomegranate in sealed PE bags at 10°C was extended up to 12 weeks with slight changes in quality [24]. The PLW of fruits cv. Jalore Seedless was up to

3.16 per cent during storage in refrigerated conditions when packed in perforated polyethylene bags [25]. The individual shrink wrapping of pomegranate cv. Ganesh held at 8°C showed the maximum shelf life of 70 days as compared to 20 days with wrapping and 15 days without wrapping at 25°C [26].

8. Post-harvest management

Pomegranate, physiologically being a non-climacteric fruit, does not ripen after harvest, and hence needs to be harvested when fully ripen on plant producing the optimal organoleptic characteristics [27, 28]. The storage life of pomegranate fruit at room temperature is merely around 10–15 days [29]. Temperature, relative humidity and atmosphere composition of gases are the environmental factors that decide the storage life of pomegranate and can be used effectively to increase the post-harvest life through the reduction in the respiration and the losses due to physiological and fungal decay [27].

However, the research efforts have helped to increase the production of pomegranate, minimize the post-harvest losses and enhance the shelf life for the purpose of obtaining maximum profit as well as the availability of fruits for longer season.

Careful post-harvest handling has been recognized as one of the important means to extend the post-harvest life of pomegranates as it reduces the majority of the mechanical damage (bruises, scrapes, cuts, compression, etc.) [27]. Modification and management of environmental conditions is another essential way to prolong the commercial life of pomegranates, and optimizing the environmental conditions to control the respiratory activity, transpiration and the development of microbial pathogens through forced air cooling to the temperature around 5°C reduces the physiological disorders and maintains the fruit quality specifications even for 2–3 months storage and extends the pomegranate commercial life [27].

The relative humidity is the second important factor and is closely related with storage temperature. To make the fruits healthy and clean, they must be gently brushed and stored with a relative humidity around 90–95 per cent to minimize the weight loss without increasing the microbial development and decay [27].

During the last few years, the activity in research and development on post-harvest management of pomegranate fruit has aimed at the application of new post-harvest storage technologies to extend the storage life of pomegranates, keeping the original quality of the freshly harvested fruits. It was targeted to the application of new refrigerated storage technologies (controlled atmospheres, modified atmosphere packaging with CO_2 enrichment and/or reduction in O_2), use of thermal treatments for fruit conditioning and curing, and intermittent warming during cold storage to avoid fungal development and the physiological disorders, physical treatments to keep the original quality of the fruit, avoiding fungal development, and the loss of quality characteristics (colour, flavour and texture) and nutritional properties (vitamins, antioxidants, health-promoting agents, etc.) having an effect on the fruit physiology and biochemistry and on the development of micro-organisms that contaminate the fruit surface [27].

8.1 Controlled and modified atmospheres

Studies in Israel had reported chilling injuries with symptoms, appearance of depression and browning of husk in post-harvest storage below 6°C. This damage caused by low temperatures can be inhibited by storing pomegranate fruits at temperatures between 2 and 6°C with 2–4 per cent O_2 [30]. Further study was conducted to observe the effect of post-harvest storage at 5°C under controlled atmospheres on pomegranate quality and incidence of chilling injury recently [31]. Different controlled atmospheres were tested, and the quality of the fruits was compared to that of the fruits stored in air. Pomegranates of cultivar 'Mollar' were stored up to 8 weeks at 5°C and under 95 per cent relative humidity and were later transferred to 20°C for 6 days to simulate the commercialization period. It was further observed that storage under controlled atmospheres (10 per cent O_2 and 5 per cent CO_2; 5 per cent O_2 and 5 per cent CO_2; 5 per cent O_2 and 0 per cent CO_2) reduced weight loss, fungal decay and chilling injuries (husk surface scald). Though these treatments were found efficient to increase the quality or shelf life of pomegranate fruits, it however reduced the vitamin C and sugar contents in the fruit juice. In case of cultivar 'Wonderful', pomegranates were stored in air enriched with 10 or 20 per cent CO_2, and it was observed that the colour of the seeds stored in air increased during refrigerated storage (10°C, 6 weeks), while the seed pigmentation was less intense in those of pomegranates stored in 10 per cent CO_2 and even decreased in those fruits stored in 20 per cent CO_2. This increased pigmentation in seeds was well correlated with the PAL activity of pomegranate seeds, a key enzyme of phenolic metabolism [32]. Nevertheless, these researchers found that storage under atmospheres with a moderate CO_2 composition (10 per cent) prolonged the storage life of pomegranates and kept the original quality, including an adequate seed colour.

Study was also conducted for the use of plastic packaging for pomegranate fruits as well as pomegranate arils, in micro-perforated polypropylene bags. It was observed in these studies that the selective permeability of the polypropylene films for different gases in the bags are hermetically sealed, and for the respiration of pomegranate fruits, a modified atmosphere is generated around the fruit, which is enriched in CO_2 and poor in O_2. So it is important to use films that allow reaching the adequate gas levels, to produce the expected beneficial effects, without triggering the fermentative metabolism that will lead to off-flavours. Thus, it is possible to prolong the storage life of the fruits with an acceptable quality, showing the fruits an exceptional appearance during the commercialization period. This modified atmosphere packaging technique has also been used for the successful storage of minimally processed pomegranate seeds [8, 33], and a patent has been developed in the post-harvest and refrigeration laboratory of the CEBAS (CSIC) Institute [27].

8.2 Thermal treatments during cold storage

Though the pomegranate fruit is cultivated in tropics and subtropical areas, where this fruit has wide harvesting period, refrigeration is the only means to prolong the storage life of fruit up to 3 months; still, studies on the storage of pomegranates under refrigeration have received little attention.

In case of Spain, pomegranate harvest for the late cultivars starts in mid-September and end in the middle of November. To extend the commercialization, pomegranates are normally refrigerated for several weeks until Christmas. However, during their storage, fruits are affected by several physiological and enzymatic disorders and fungal attacks, deteriorating their quality. Spanish 'sweet' pomegranates suffer chilling injury if stored for more than 2 months at temperatures below 5°C, exhibiting symptoms, viz. browning of the rind, pitting, scald, an increased sensitivity to fungal development and decay, and internal discolouration and browning of the seeds [34]. Some cultivars, i.e. 'Wonderful', can be stored without problems for 2 months at 5°C, and the minimal safe temperature to store sweet pomegranates is 10°C [35]. This temperature, however, does not prevent fungal development. Previous studies have demonstrated that though the conventional storage of 'Mollar' pomegranate at 5°C and 90–96 per cent RH up to 8 weeks leads to acceptable decrease in fungal decay losses, the risk of chilling injury was not completely prevented. Storage under controlled atmospheres (5 per cent CO_2 and 5 per cent O_2) at 5°C and above 95 per cent relative humidity for 2 months improved the quality attributes of the freshly harvested pomegranate, although a moderate husk scald was observed [31].

A technique of intermittent warming that has proved useful in the prevention of chilling injury symptoms in other products has been recently applied to pomegranates [36]. Under this treatment, 'Mollar de Elche' pomegranates were stored at 0°C and 5°C and 95 per cent RH for 80 days with intermittent warming treatments at 20°C in cycles, for 1 day in every 6 days storage, followed by a commercialization period of 7 days at 15°C and 70 per cent RH. It was observed that while intermittent warming during the storage at 0°C prevented fungal development, increased susceptibility to chilling injury symptoms was there and storage at 5°C reduced considerably the chilling injury symptoms, but fungal attack was not completely inhibited. The warming treatments led to very good results in keeping the quality of pomegranates, especially when they were applied to pomegranates stored at 0°C.

Pre-conditioning at moderate temperature (30–40°C) and high RH (90–95 per cent) for a short period of time (1–4 days) is another technological treatment that has been utilized with success in pomegranate. This technique is also known as curing, that is applied before the conventional refrigerated storage. As reported, a pre-treatment at 35°C and 90–95 per cent RH applied for 3 days prior to a refrigerated storage for 80 days at 5 or 2°C and 90–95 per cent RH has reduced the pitting and husk superficial scald considerably (produced by the enzyme polyphenol oxidase) as compared to control pomegranates without the pre-conditioning treatment. The effects observed were more marked when the conservation was carried out at 2°C than at 5°C, particularly during the additional period of 1 week at 15–20°C and 70–75 per cent RH, applied to simulate the retail sale period [27, 36].

Pre-harvest treatment of 'Mollar de Elche' pomegranate with salicylates, especially salicylic acid @ 10 mM concentration, is found to be a safe, natural, and new tool to improve fruit quality parameters such as firmness, aril colour and individual sugar, especially fructose and organic acid contents at harvest. Moreover, aril content on bioactive compounds, such as phenolics, anthocyanins, and ascorbic acid, was also increased by salicylate treatments at harvest and during prolonged cold storage at 10°°C and 85–90 per cent of relative humidity also recording the reduction in weight loss. The quality traits and the concentration of bioactive compounds in fruits were found to be maintained at higher levels during the cold storage [18].

A sharp increase in PLW of fruits stored at room temperature is observed, whereas the increase in PLW is very slow in fruits stored in cool storage. The juice content of pomegranate fruit decreased significantly with the increase in storage period irrespective of post-harvest treatments and storage conditions, but the rate of decrease was faster at RT than in cool storage. Treatment of coating the fruits with wax coupled with 0.1 per cent of carbendazim recorded very low PLW as compared to other treatments under both the storage conditions. This might be due to the fact that wax acted as a barrier for the loss of moisture from the fruit surface and found to be more effective in high RH and low-temperature prevailing in cool storage. The fruits treated with wax showed low reduction in juice content as compared to control. Wax coating and fungicides when applied to fruit showed the moisture loss and respiration by forming a film around the fruit thereby retaining juice percentage. The shelf life of the fruits could be extended up to 30 days under RT and up to 65 days under control storage when treated with wax (9 per cent) coupled with 0.1 per cent carbendazim. The fruits stored in cold storage were more fresh, firm and glossy in appearance and attractive as compared to those stored at room temperature. The organoleptic rating of the fruits in terms of colour, flavour and texture was maximum in the fruits treated with wax coupled with 0.1 per cent carbendazim [29].

9. Pomegranate processing

Pomegranate indicates the great scope for the processing into value-added products having extended shelf life (**Figure 1**). Well-matured big size fruits with attractive colour are readily picked by consumers, whereas low-grade pomegranates with fruit disorders, such as sunburnt husks, splits and cracks, reduce marketability and consumer acceptance. The new sector of pomegranate processing allows the use of such low-quality fruits that cannot be commercialized for the preparation of the new products. Though there is a great potential for pomegranate-derived products, the industrial processing as well as consumption of pomegranate is scarce due to peeling difficulties and lack of technologies for industrial processing of pomegranate [33, 38].

10. Pomegranate juice processing

Considered as a whole, pomegranate contains 48 to 52 per cent of edible part, which comprises 78 to 80 per cent juice and 22 to 20 per cent seed. Juice is extracted

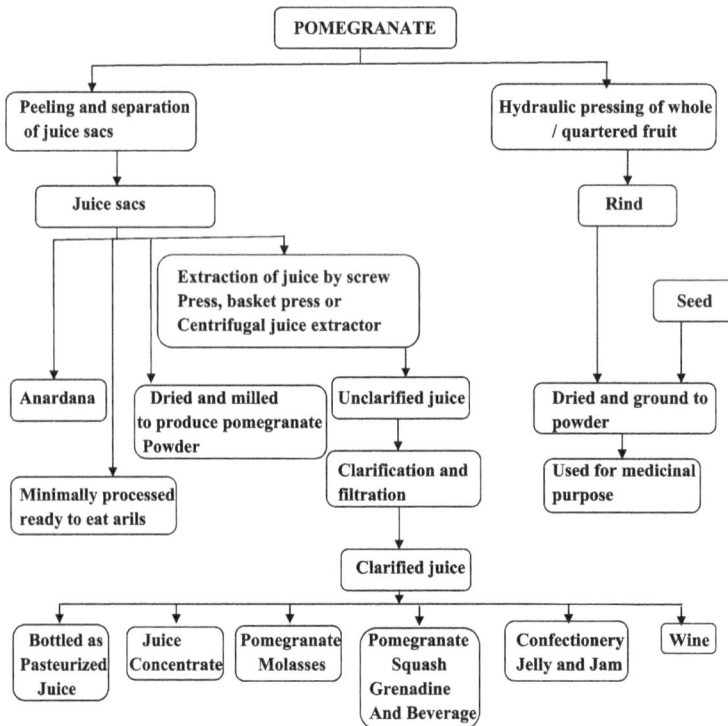

Figure 1.
Utilization of pomegranate for industrial processing: a flow chart (source: Yadav et al. [37]).

by crushing seeds along with arils and marketed as a fresh juice as it has excellent flavour, attractive fragrance, delicious taste and high nutritive and medicinal value. Juice extraction is one of the important methods of value addition, and this juice can be processed further to prepare squash, syrup, nectar, jelly, concentrate and such other products (**Figure 2**). Also, pomegranate juice is used as an ingredient providing colour to the other products.

Pomegranate contains punicalagins, hydrolysable tannins, anthocyanins and ellagic acid, and the compounds responsible for antioxidant capacity. Hence, pomegranate juice has higher contents of antioxidant compounds than the fresh fruit. While extracting the juice, relevant changes on the bioactive compounds are observed, in which punicalagins increase while ellagic acid and anthocyanins decrease. The flavour of pomegranate juice is characterized by esters, alcohols, and terpenes [40]. Terpenes are the predominant group in the pomegranate fresh juice aroma, while furans are only present in processed juices.

The phenolic constituents of pomegranate such as the anthocyanins give the colour, and other polyphenols such as flavonoids and some non-flavonoids are

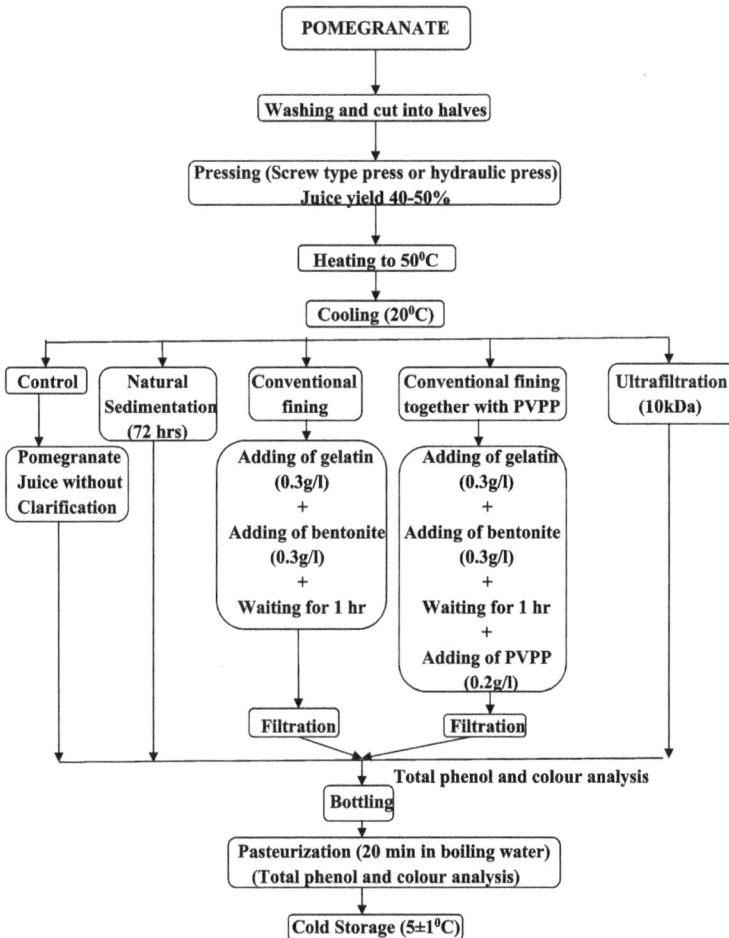

Figure 2.
Flow diagram of pomegranate juice processing technology: use of different clarification techniques (source: Alper et al. [39]).

responsible for antioxidant properties, astringency and bitterness to juice [41]. The pomegranate juice is a rich source of polyphenols. The antioxidant qualities of pomegranate juice make it appealing for the production of health supplements and nutraceuticals [42].

10.1 Extraction of pomegranate juice

In case of the pomegranate, the foremost challenge in juice extraction is the peeling of the fruit as it is time-consuming and irritating as the hands get stained due to polyphenols and oxidative enzymes. To this date, commercial pomegranate juice production has only been mentioned by Cemeroglu [43], Vardin [44] and Saxena et al. [2].

For juice extraction, the fruits are prepared by rolling it on the hard surface to weaken the seed sacks within the fruit. MPKV, Rahuri, had developed a prototype machine which can separate arils and skin from fruits without causing damage to arils. CIPHET, Ludhiana, India had designed a hand tool for easy extraction of arils from fruits. Cut opening of the fruit, seed separation and pressing in screw press or basket press are the basic methods for the extraction of juice. In another method, the fruits are quartered and crushed or the whole fruits are pressed in hydraulic press and juice is strained out. Juice is extracted from the mature pomegranate seeds of sweet-type cultivars like Mollar and Bhagwa by pressing and liquefying. The pressing method gave the greater quantity as well as better quality. On whole fruit basis the juice yield is about 42 per cent, while on aril weight basis the yield is about 70 per cent [45]. Packaged press method for the extraction of pomegranate juice from separated arils, clarification of the pomegranate juice with gelatin at 4°C overnight and storing filtered juice at −30°C are described by Ozkan [46]. While, Saxena et al. [2] extracted 36.41 per cent of the juice by cutting the pomegranate fruits into quarters and pressing them in a rack and cloth hydraulic press under moderate pressure. The hydraulic extraction of juice should be at a pressure less than 100 psi to avoid undue yield of tannins from the rind. The phenolic constituents of the pomegranate juice are responsible for the colour, astringency, bitterness as well as the formation of cloudy appearance of fruit juices during concentration and storage as reported by de Simon et al. [47] and Spanos and Wrolstad [48]. Adsule and Patil [49] reported that average yield of juice was about 50 per cent on whole fruit basis, while from grains the yield was about 76 to 85 per cent by hand-press method. Chobe [50] separated grains and rind by pomegranate seed extractor and extracted 55 per cent pomegranate juice by crushing them in screw-type juice extractor. Vardin and Fenercioglu [51] obtained the 30 to 40 per cent pomegranate juice by pressing the whole pomegranate fruit in the manually operated packaged-type press in batches of 10 kg for 5 min. They also reported that the pomegranates can be pressed as a whole or as a divided or as granulated sac, but the best results were obtained by using whole fruits considering the time and cost, but the juicer with pressure of above 10 kg/cm^2 resulted in the juice with excessive astringency and bitterness due to squeezing of phenolic compounds. Neifar et al. [52] extracted pomegranate juice with pH 4.1 and TSS 15^0 Brix by using the Philips centrifugal electric juicer (centrifugal method) and stored the juice at −20°C. Alpher et al. [39] described the process of extraction of pomegranate juice by pressing the two halved cut fruit into laboratory-type press which yielded 40–50 per cent juice. Marti et al. [53] described a process for the extraction of pomegranate juice having 3.8 pH in which aril pressurization followed by centrifugation at 6000 rpm for 10 min was used. Kumbhar et al. [54] reported that the average pomegranate juice extractions on whole fruit basis by hand-press method, mechanical method and screw-type juice extractor method were 71.6, 66.0 and 81.0 per cent, respectively.

The pomegranate juice can also be extracted by squeezing the arils gently by hand-press method [55, 56]. However, Catania et al. [57], in industrial juice extraction with hydraulic press and a pneumatic press, recorded the better qualitative results in pneumatic press, whereas Catania et al. [58] reported that the application of different pressure values during pomegranate juice extraction process allowed the products of different quality.

10.2 Clarification of pomegranate juice

After the extraction, next important step in fruit juice processing is clarification or fining. Clarification helps to remove active haze precursors and thus decreases the potential for haze formation during storage. If the whole fruit is crushed or pressed with excessive pressure, a very large amount of excessively 'puckery' tannin, which is present in the rind of the pomegranate fruit, enters into juice and makes it undrinkable [44]. Nutritionists, however, recommend in contrast to preserve these compounds during the fruit juice processing because of their health protective effects. Pomegranate juice contains only trace amount of pectin that can be filtered easily after pressing without clarification. However, clarification is necessary to prevent the formation of cloudy appearance during storage and also to improve the taste of the product. Without clarification, the product has a bitter taste due to high tannin content. These polyphenols contribute to the haze formation through the mechanisms involving prior polymerization or condensation leading to the formation of the polymeric complexes and are collected at the bottom of the fruit juice bottle when stored. The main purpose of the clarification is to reduce the amount of the tannin and decrease the astringency of the product. Effective use of clarification agents requires optimization of their method of preparation as well as determination of the appropriate concentration needed to achieve clarification. For clarification, gelatin, bentonite, clays, etc., may be used as flocculating agent. Sahin et al. [59] observed the positive charge of gelatine at the pH of fruit juice which removed the negatively charged phenolic compounds by forming heavy flocculent precipitate forms. But if the gelatin is added above the level, turbidity increases [60]. Pectinase enzyme plays important role in clarification of the fruit juices by depectination [61]. The centrifugation method may also be employed for the clarification of the fruit juices. Vardin [44] reported the conventional heating treatment to raw pomegranate juice to inactivate naturally present enzymes and to destroy the vegetative micro-organisms. He further opined that heating or pasteurization of the pomegranate juice can be applied after clarification and filtration as heating before clarification increases the stability of haze formation which hinders the clarification of juice permanently. The natural clarification can be suitably employed for the clarification of pomegranate juice because the pomegranate juice was quite resistant to microbial spoilage at refrigeration temperature due to the presence of the polyphenolic compounds [43]. The most effective method to remove the phenolic compounds in pomegranate juice was the conventional fining with gelatin (300 mg/l) and bentonite (300 mg/l) along with polyvinylpolypyrrolidone (PVPP) [39] (**Figure 2**). Vardin and Fenercioglu [62] clarified the basket-pressed juice with gelatin, polyvinylpolypyrrolidone (PVPP) and natural sedimentation and reported that the phenolic substances were controlled in each clarification method. The most effective method of clarification was the application of 1 g/l gelatin before the heat treatment as it reduced the phenolic substances to an acceptable level, decreased turbidity, and preserved anthocyanins and colour density. In order to reduce the amount of tannin in pomegranate juice, Bayindirli et al. [63] found that the addition of 2 g/l gelatine is the most effective method of clarification which resulted in the clear and rich coloured juice as compared to the natural clarification which gave turbid juice. Removal of phenols can be accomplished with the help of filtration methods such as ultrafiltration. Neifar et al. [52] investigated that the laccase enzyme application to pomegranate juice resulted in the 40 per cent reduction of the total phenol but induced a threefold decrease of juice clarity. This drawback was overcome by the ultrafiltration of laccase-treated juice giving clear and stable pomegranate juice. Pectinases play a crucial role in clarification, extraction, reduction in viscosity, removal of peels and increment of the yield of fruit juices. The use of pectinase enzyme for the clarification of the fruit juices by depectination is reported by many authors [61, 64].

Ashima Kapoor and Hina Iqbal [65] used the tannase produced from fungal strain of *Trichoderma harzianum* for pomegranate juice clarification at 40°C and noted the tannin reduction of 57 per cent without the loss of its biochemical attributes such as pH, viscosity and sugar content and protein content.

10.3 Packaging of pomegranate juice

Selection of packaging material as well as processing influence the quality of foods, altering colour and nutrient composition during storage as a result of contact with oxygen and light transmission through them. Paperboard cartons with low-density polyethylene (LDPE) coating or glass containers are commonly used materials for juice packaging. Oxygen and light have destructive effects on the anthocyanin during storage. So, the packaging material also plays an important role in the colour stability of stored pomegranate product. Perez-Vicente et al. [66] assessed the influence of packaging material on colour and bioactive compounds of pasteurized pomegranate juice during storage at 24/18°C and 40–-50 per cent RH. They opined that the organoleptic quality of juice could be altered by packaging material, even if nutritional quality is not influenced suggesting that the oxygen permeability of the packaging material (which is the more damaging factor than light for pomegranate juice) should be minimized to avoid the detrimental effects on the retention of colour and some bioactive compounds. Glass containers were found to be better as compared to high-density polyethylene or polyvinylchloride containers with regard to retention of anthocyanins, vitamin C and organoleptic quality of the fruit juices [67]. Wasker and Deshmukh [68] studied the effect of light on the quality of stored pomegranate juice. The results showed that pomegranate juice packed in amber-coloured glass bottle retained better anthocyanins as compared to juice packed in colourless bottles.

Fathy [69] evaluated four different packaging materials for pomegranate juice during 8 weeks of storage at 4°C, the results of which approved that cartons consisting outside the polyethylene layer 13.5 per cent + cardboard 27 per cent + an inner layer of 59.5 per cent aluminium foil (Tetra Pak) as superior packages than cartons with an inner layer of ethylene vinyl alcohol copolymers and transparent PET bottles. It retained the high quality of pomegranate juice, preserving the high vitamin C content, intense red colour, fresh pomegranate flavour and absence of off-flavours. It also leads to the lower oxygen content in the headspace of the containers.

10.4 Storage of pomegranate juice

Pomegranate juice contains coloured pigments like anthocyanin, and the stability of these pigments is dependent on number of factors, viz. temperature, oxygen, light, pH and enzymatic action. Among these factors, storage temperature is considered as the important one. It has been reported by many researchers that pomegranate juice can be stored at ambient temperature as well as in cold storage at around 5 ± 1°C. But as compared to storage at ambient temperature, better retention of anthocyanins and reduction in enzymatic activity were reported in pomegranate juice stored in cold storage. As reported by Ahire [70], higher acceptable organoleptic score was observed in the hand-pressed or mechanical-pressed pomegranate juice packed in the glass bottles which was stored satisfactorily up to 3 months under cold storage (5 ± 1°C) conditions. Suryawanshi et al. [71] reported that the pomegranate juice can be stored up to 60 days at room temperature with the minimum changes in TSS, acidity, pH, total sugar, reducing sugars and tannin when pasteurized at 70°C and added 500 ppm sodium benzoate. Adam and Ongley [72] reviewed the beneficial influence of low-temperature storage on various pigmented fruit products, and they observed that bottling of fruits at low pH (between 1 and 2), without adding sugar, led to small but significant increase in the pigment stability. Pigment degradation was faster in juice stored at ambient (35°C) temperature. Shelar [73] and Sandhan [55] reported an increasing trend in TSS, pH, reducing sugars, non-reducing sugars and total sugars during 3 months storage of pomegranate juice-based carbonated beverage both at ambient and in cold temperature storage.

11. Minimally processed (ready-to-eat) fresh pomegranate arils

As it has been discussed earlier though the pomegranate fruit is rich in nutrients and antioxidants, the preparation of the arils is tedious, difficult and time-consuming procedure, and it makes the pomegranate fruit unpopular as a table fruit. It is for this reason the challenge of the development of 'ready-to-eat' pomegranate arils has been approached by several research groups in Spain and the USA as reported by Artes et al. [74]. In recent years, minimally processed 'ready-to-eat' pomegranate arils have become popular due to their convenience, high value, unique sensory characteristics and health benefits (**Figure 3**). MAP of minimally processed pomegranate arils offers additional tool for optimal use and value addition and also for the utilization of the lower-grade fruits, with increasing global interest in nutritional value as well as post-harvest handling of pomegranate fruit as reported by James Caleb et al. [76]. Minimally processed arils easily deteriorate in texture, colour, overall quality and a reduction in shelf life; hence, maintaining the nutritional and microbial quality of pomegranate arils is a major challenge [8, 33]. Washing with the sanitizing agents to reduce the initial microbial load, pH modifications, use of antioxidants, modified atmosphere packaging and temperature control are certain components of minimal processing as reported by Sepulveda et al. [75] (**Figure 3**). Washing the arils with the chlorine solution, followed by a mixture of ascorbic and citric acids and storing the seeds at 1°C in polypropylene films for the arils of the cultivar 'Mollar de Eche', allowed the formation of a modified atmosphere appropriate for the conservation of these arils. The preparation of the arils under very clean conditions and at temperatures close to 0°C prolonged the life of this product and maintained its quality. Storage at the higher temperatures (4–8°C) produced the product with lower quality and a shorter commercial life [8]. Potential increase in the shelf life of pomegranate arils by ensuring the microbial safety and monitoring the storage temperature with TTI is offered by the novel technologies such as smart packaging. Use of natural or non-destructive products as preservatives such as honey and UV-C radiation should be done in combination with MAP [76]. Ayhan and Esturk [77] found that the though pomegranate arils packed with air, nitrogen and enriched oxygen kept acceptable quality attributes on 18th day, the marketability period was limited to the 15th day for the low oxygen atmosphere. The effect of harvest time, use of different UV-C radiation and passive

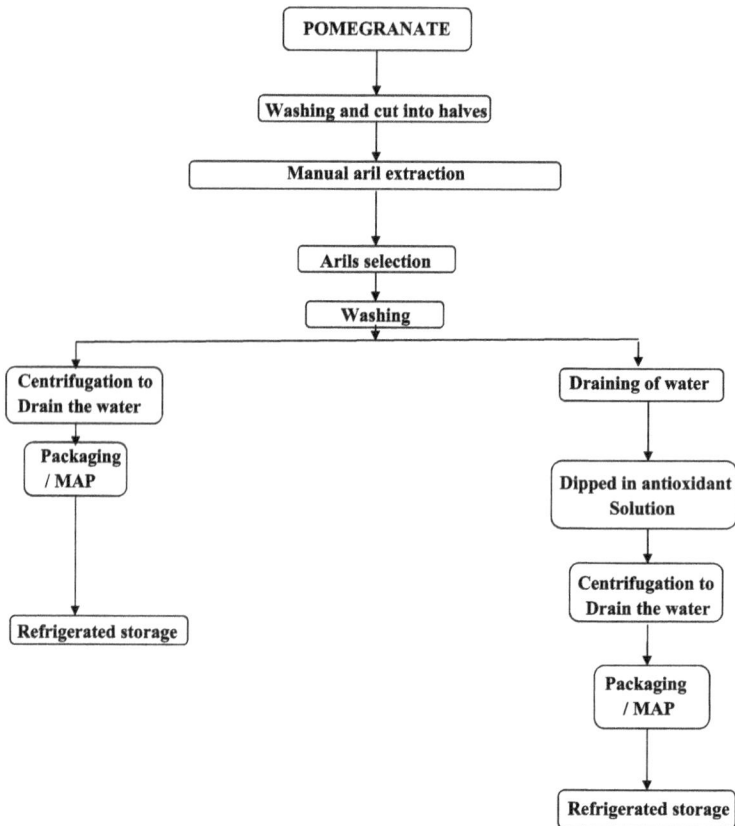

Figure 3.
Flow diagram of minimal processing of pomegranate arils (source: Sepulveda, et al. [75]).

MAP with polypropylene basket sealed with BOPP film storage on sensory, chemical and microbial quality as well as on the shelf life of minimally fresh processed pomegranate var. Mollar de Elche arils was studied by Lopez-Rubira et al. [38]. The rate of respiration of fresh processed arils was higher in the post-harvest than in the earlier harvested fruits. No significant differences were observed between the control and UV-C-treated arils, and there was no observable interaction between passive MAP and UV-C treatment, except that the CO_2 accumulation within aril packages was higher in December harvest than those of October due to their respiration rate. They also reported that unclear results were obtained due to the effect of UV-C radiation on the microbial growth of pomegranate arils and the shelf life of commercially produced pomegranate arils can be increased up to 10 days with the use of 100 per cent nitrogen in pet packages. No significant change in the total anthocyanin content of arils was observed. They also reported that the microbial counts of minimally fresh processed arils increased the shelf life, with mesophilic counts of control arils processed in October slightly higher than those from December. A low count of micro-aerophilic lactic acid bacteria (LAB) after 10 days of aril storage without any trace of fermentative metabolism was also recorded.

Nunes et al. [78] in their investigation regarding metabolic response of the minimally processed pomegranate arils cv. Hicanar to UV-C treatments and stored at 2°C and 6°C observed that the UV-C illumination had an effect on the increase of phenol content in pomegranate arils, but TSS and citric acid percentage remained unaffected. Similarly, Maghoumy et al. [79] recorded that the combination of UV-C and high oxygen (initial 90 kPa O_2) active MAP treatments for cold storage at 5°C and 90 per cent relative humidity as a good option to improve and extend the shelf life of fresh-cut arils as it preserved the superoxide dismutase and catalase had lower values of peroxidase and polyphenol oxidase and maintained the concentration of antioxidant compounds up to 14 days.

Gil et al. [33] in their study on pomegranate variety Mollar reported that arils washed with chlorinated water (100 mg/kg) and antioxidants solution (5 g/l ascorbic acid and 5 g/l citric acid), packed in OPP film, using an initial atmosphere actively modified to 0 ml/l CO_2 and 20 ml/l O_2 and stored at 1°C, can be stored up to 7 days maintaining a good quality and appearance, without visible attack of moulds or off-flavour developments. Similarly, Sepulveda et al. [80] investigated that pomegranate arils cv. Espanola washed and immersed in antioxidant solution (sodium hypochlorite, 200 ppm, 5 per cent ascorbic acid and 5 per cent citric acid) for 1 minute and stored at 5 ± 0.5°C in three semipermeable (SP) packaging can be stored for 7 days with commercial marketable acceptance. The best ability to maintain the physical, chemical and microbiological characteristics of the minimally processed pomegranate arils was shown by BB4 bags, but the sensory characteristics decreased from 7 to 14 days storage. Palma et al. [81] during their studies on the chemical and organoleptic characteristics of minimally processed seeds of pomegranate cv. Primosole, packed in a 40-μm-thick polypropylene film, stored at 5°C for 10 days observed that a passive modified atmosphere was established within the package, with a progressive increase in CO_2 and decrease in O_2 level, with increased ethylene concentration at the end of the storage. No significant changes in chemical properties were recorded, but increase in titratable acidity was observed in packaged seeds.

Palma et al. [82] despite physiological disorders as a severe external chilling damage appearing during storage on the rind made most of the fruit not marketable stored for 30 or 60 days at 5°C and 90 per cent RH, the overall quality with titratable acidity, total soluble solids, total polyphenol and anthocyanin content, antioxidant activity, sugar content and juice colour of ready-to-eat arils of 'Primosole' pomegranate processed at harvest and after 30 or 60 days of storage of whole fruit did not change substantially at the end of each storage time over fresh whole fruit. Similarly, the behaviour of ready-to-eat arils packaged in polypropylene trays wrapped with a polypropylene film to generate a passive modified atmosphere and stored at 5°C and 90 per cent RH for 10 days did not show marked differences in chemical and physical parameters giving the opportunity to store pomegranates fruit intended to be processed as ready-to-eat arils for a longer time.

Hess-Pierce and Kader [83] concluded that arils of '*Wonderful*' pomegranate can be stored without changes in the physical and chemical characteristics of the fruit up to 16 days at 5°C with 20 per cent gas composition of CO_2. They also observed the susceptibility of the mechanically damaged arils to moulds after 12 days. Ayhan and Esturk [77] recorded the commercially acceptable shelf life of minimally processed pomegranate cv. Hicaznar arils treated in a solution of 1 per cent citric acid (w/v) and 100 μLL − 1 chlorinated water as 18 days under air, nitrogen (100 per cent) and enriched oxygen (70 per cent) atmospheres, while 15 days under the low oxygen (5 per cent) atmosphere with package type of PP tray with BOPP film at 5°C storage. They detected the slight or no significant changes in chemical, physical and sensory quality during refrigerated storage. However, Belay et al. [84] reported that at both 5°C and ambient storage, arils of 'Wonderful' pomegranates under super atmospheric O_2 significantly maintained the lowest microbial counts, better colour and texture properties compared to other MA conditions and also promoted the production of highest composition and amount of volatile compounds essential for flavour profile and antimicrobial effects identified in pomegranate arils. In Belay et al. [85] reported that bi-axial-oriented polypropylene-based film (PropaFilm) as a packaging film (66 per cent) fitted with 100 per cent cellulose-based film (NatureFlex) window was the best to maintain the quality of pomegranate arils for 9 days at 10°C temperature and 91 ± 2 per cent RH storage conditions and regulating the in-package relative humidity.

The effects of CO_2 and O_2 on the respiration rate of pomegranate arils cv. Hicaz evaluated the suitability of the common packaging material for designing the MAP was studied by Ersan et al. [86]. Respiration rate of the arils was significantly affected by O_2 and CO_2 concentrations in the surrounding atmosphere. Though lower respiration rate was achieved in 2 per cent O_2 with 10 or 20 per cent CO_2, MAP design analysis indicated that these target levels of O_2 and CO_2 cannot be achieved by LDPE and PP materials for commercial packages. The storage atmosphere enriched in CO_2 (10, 15 and 20 per cent) helped to prolong the commercial life of pomegranate seeds up to 8 days at 10°C and 12 days at 5°C [27].

Adiletta et al. [87] reported semipermeable (SP) packaging system as a valid tool to preserve the qualitative and nutraceutical values in ready-to-eat pomegranate arils of cv. 'Purple Queen' over micro-perforated (MPP) system during cold storage at 5°C for 16 days and suggested that SP packaging system for ready-to-eat arils could be used in food industry applications as a convenient alternative to fresh fruit consumption. The SP packaging system found to reduce oxidative stress that increases antioxidant enzyme activities such as superoxide dismutase, catalase, and ascorbate peroxidase involved in the first line-of-defence against reactive oxygen species, detoxification, and the inhibition of PPO and GPX activity that are involved in aril-browning processes. SP packaging prevented the microbial growth and allowed the storage up to 16 days without exceeding the maximum-acceptable microbial limit than the MPP.

Honey has been used since ancient times as a sweetening agent in food and is the only concentrated form of sugar available worldwide. The use of honey treatments has been explored in preserving the fresh-like quality of arils and other cut fruits and to extend their shelf life. Efficacy of varying concentration of 10 and 20 per cent honey dip treatment on the quality and shelf life of minimally processed pomegranate arils cv. Hicaznar stored at 4°C in loosely closed containers was evaluated by Ergun and Ergun [88]. It was reported that honey solution (10 and 20 per cent) dipping 5 min can help in storage of arils at 4°C for 10 days with brilliant aroma than those treated with water. This treatment also helped in lowering the softening of the arils. The total aerobic count was lower in honey-treated arils, and the counts increased across all the treatments compared with the count immediately after the treatment. It also extended the quality of arils by delaying quality loss, microbial development and pigment changes, thus providing the safe organic method.

Martínez-Romero et al. [89] washed fresh pomegranate arils of cv. Mollar de Elche in a solution containing 100 μLL^{-1} chlorine (NaOCl) for 5 min, followed by further rinsing in tap water and removal of excess water. It was followed by post-harvest treatments prior to storage in rigid polypropylene boxes for 12 days at 3°C. Among them, the combination of *Aloe vera* gel at 100 per cent + ascorbic acid and citric acid at 1 per cent retained the quality parameters of minimally processed arils such as firmness, colour and bioactive compounds along with reduction in microbial spoilage (significantly lower counts for both mesophilic aerobics and yeast and moulds) and increased the levels of total anthocyanins and total phenolics. Sensory analysis scores for flavour, texture, aroma, colour and purchase decision were higher in arils under this treatment without any off-flavours in pomegranate arils perceived as a consequence of *A. vera* gel treatment.

Ashtari et al. [90] reported that the gamma irradiation at 1–5 kGy reduced the population of microbial agents including bacteria, fungi and yeasts, but at the cost of reduction in amount of polyphenols, ascorbic acid, anthocyanin and antioxidant capacity. An increase in the amount of oxidative hydrogen peroxide in pomegranate aril and quality reduction in aril at low dose of 1 kGy was not remarkable and it also prevented the browning of the arils through the control of PPO enzyme activity, hence application of gamma irradiation for increasing the pomegranate aril shelf life is recommended.

Ozdemir and Gokmen [91] recommended coating of the pomegranate arils with an aqueous mixture of 1 per cent chitosan and 1 per cent ascorbic acid (chitosan-ascorbic coating) that significantly prolonged the lag time of micro-organisms and extended the microbiological shelf life of arils up to 21 days during storage at 5°C owing to antimicrobial activity of chitosan along with high sensory scores (colour, taste and aroma) which were found quite acceptable even after 25 days of refrigerated storage. Visual quality pomegranate aril in this treatment was protected due to ascorbic acid in the binary coating mixture as it is very effective anti-browning agent. Hasheminejad and Khodaiyan [92] studied the effect of four different coating dispersions, viz. chitosan, clove essential oil, chitosan nanoparticles and clove essential oil-loaded chitosan nanoparticles (CEO-ChNPs) on shelf life, and the quality of minimally processed pomegranate arils during cold storage at 5°C.

Type of product	Package film	MA composition		Storage temp. (°C)	Storage period	References
		%O$_2$	% CO$_2$			
Pomegranate arils cv. Mollar de Elche	Oriented polypropylene (OPP)	188 ml/l OPP-CO$_2$ 206 ml/l OPP-N$_2$ 203 ml/l	22 ml/l 3 ml/l 4 ml/l	1	7 days	[8]
Pomegranate arils cv. Mollar	Semi-permeable plastic bag	1	30	4	10 days	[93]
Pomegranate arils cv. Mollar de Elche	OPP, 40 μm thickness	12.5 13.5 18.5	8.5 7.5 2.5	8 4 1	7 days	[8]
Pomegranate arils cv. Mollar de Elche	Polypropylene basket sealed with BOPP (October) Polypropylene basket sealed with BOPP (December)	2–5 kPa 2–5 kPa	20.1– 21.6 kPa 26.9– 29.9 kPa	5	15 days	[38]
Pomegranate arils cv. Wonderful	BB4 (Cryovac based on ethyl vinyl acetate) BE (Cryovac based on ethyl vinyl acetate) Perforated polyethylene bags	1 12 Not reported	22 2	4	14 days	[75]
Pomegranate arils cv. Primosole	Polypropylene	6.5	11.4	5	10 days	[81]

Source: James Celeb et al. [76].

Table 4.
Summary of modified atmosphere packaging of pomegranate arils, different cultivars, types of packaging film, modified atmosphere composition with temperature and duration of storage under MAP.

The CEO-ChNPs significantly extended aril shelf life for 54 days due to the controlled release of the preserved volatile oil from ChNPs and their own inhibitory effects which significantly maintained microbial quality, weight, total soluble solid, titratable acidity, pH, total phenol and total anthocyanin content, as well as antioxidant activity and sensory quality in pomegranate arils, while uncoated arils became unusable at day 18 due to the incidence of fungal decay.

A summary of different cultivars, MAP of pomegranate arils, the types of the packaging adopted and the modified atmosphere condition attained in the packages is reviewed in **Table 4**.

12. Frozen arils

The arils were packed in polyethylene bags with syrup of 15^0 Brix and then frozen in a chest freezer, after preparing the arils same as that of minimally processed arils. The syrup concentration is similar to that of the arils juice concentration. The juice content of aril enters the syrup during freezing. Similarly, freezing of the arils coated with the sugar was also reported by Maestre et al. [94]. Due to the effect of anthocyanins, the coated sugar turned red during freezing and storage. To avoid an excessive loss of turgidity, arils should be eaten frozen.

13. Appertised arils

Appertised arils are prepared by putting the arils in syrup of 15^0 Brix and packed into metal tin. To sterilize, the tins were sealed and heated for 10 min. Most of the tins had arils that were too soft and that tasted cooked after stabilization, and the tins which were prone to less severe heat treatment had adequate textured arils with good taste.

14. Arils in vinegar

The arils were preserved in vinegar with an acidity of 5 per cent and packed in jar, and the resultant product was the brown coloured pomegranate arils.

15. Pomegranate jam and preserves

Pomegranate juice is concentrated and mixture is heated on slow fire for long period, and the resultant product is known as *Anar Rub*, which has fairly good keeping quality. The finished product has a thick consistency and TSS content is about 70–75 per cent which can be stored for 1 year and is utilized as jam [95]. Maestre et al. [94] reported that during the processing treatment 25 per cent of the pigments are destroyed in the jams and preserves made from the frozen Mollar pomegranate juice by adding pectins, saccharose and citric acid. They continue to degrade during preservation depending on the temperature than on light. The best preservation temperature reported was 5°C.

16. Jellies

Adsule et al. [96] prepared good pomegranate jelly on a small scale from Ganesh cultivar of pomegranate. An attractive jelly can be prepared from pomegranate juice [45]. But while making jellies, approximately 50 per cent of the anthocyanins present in pomegranate juice were lost. Maestre et al. [94] investigated that the acidification of juice produced a noteworthy improvement in the colour of jelly, both initially and during storage. During storage, certain colour differences were observed, which indicates that the pH was not only the parameter responsible for this characteristic.

17. Anardana

Sour wild types of pomegranate are utilized to prepare 'Anardana'. It is dried aril of pomegranate mainly used as an acidulent instead of tamarind or dried green mango (Amchur) in North India in Indian cuisines like curries, chutney and other culinary preparations. It has anti-inflammatory properties and is a good source of dietary fibres. It is also used in the preparation of digestive candies and used in the traditional system of Ayurvedic and Unani medicine for ulcerative colitis. The improved processing technique for *anardana* consists of pre-cleaning, mechanized extraction of arils, solar/sundrying and packaging. After treating with sodium benzoate (600 ppm) for 10 min, arils are dehydrated in a drier at 45°C for 48 hours to get 10–12 per cent moisture content. It has attractive brown colour and can be stored for a long time in glass jars [97].

18. Pomegranate molasses

Pomegranate molasses is traditional Middle Eastern ingredient made from cooked pomegranate juice. It is thick and syrupy in texture, provides tangy flavour and is dark in colour. The sweetness comes from the concentration of natural sugars of fruits [37]. To make molasses, pomegranate juice is heated in a pan for about 45 min, allowing it to thicken but not overcooked. The product can be stored in air-tight container under refrigerated conditions for 3 months and typically used to flavour chutneys, curries and salad dressings to glaze or tenderize meat products.

19. Pomegranate concentrate

Pomegranate concentrate is prepared from natural pomegranate juice and is generally free from added sugar or preservatives. The commercially available juice concentrate generally contains 65–70 per cent TSS and has pH of 2.7 to 3.1. Majority of manufacturers in international market promote pomegranate juice as a health drink advocating its medicinal and health benefits. Different processes for the preparation of pomegranate concentrate are described by Maskan [98], which include keeping 60.5 ^{0} Brix by using microwave, rotary vacuum and atmospheric heating processes for 23, 108 and 190 min, respectively. Increase in the reducing sugars, glucose and fructose level to 46.46 per cent, 23.89 per cent, and 22.53 per cent, respectively, during the conventional method of pomegranate juice concentration process was recorded by Hulya Orak [99]. Also, the potassium and magnesium mineral contents of concentrate increased during concentration.

20. Pomegranate syrup/grenadine

Pomegranate syrup is used for flavouring in alcoholic drinks, soft drinks and confectionaries. It is sold commercially as grenadine. Grenadine is light pomegranate syrup prepared by mixing juice and sugar. A syrup has bright purplish red colour and delightful taste and flavour. It has about 60^0 Brix with an added acidity of 1.5 per cent with citric acid. Grenadine can be preserved by pasteurization or by the addition of sodium benzoate and stored in amber glass bottles at room or refrigerated storage to retain its quality parameters.

Thakur et al. [100] prepared the arils-in-syrup product from wild pomegranate with 10 combinations of arils, syrup and TSS of syrup of which product with 40 per cent arils and 60 per cent syrup of 75^0B was adjudged to be the best according to its sensory and some physico-chemical characteristics like colour, titratable acidity and ascorbic acid.

21. Pomegranate beverages

Pomegranate juice can be processed for RTS beverages, blend and squash. These products are with delicate flavour. Ready-to-serve beverages and beverage blends are with 15 per cent juice content, and squash is prepared with 40 per cent juice. Squash is served in ratio of 1:4 with water (one part of squash with four parts of water).

The refrigerated probiotic beverage was found more stable for microbial, physico-chemical and colour attributes during storage, whereas beverage kept at room temperature could be microbiologically safe for 1 week [101].

The low alcohol beverage could be prepared from the fermentation of pomegranate juice as single substrate (100 per cent) or mixed with orange juice (1:1) using kefir grains which was proposed for a low-alcohol beverage production with high nutritional value deriving from both the substrates and the culture used and were also having the beneficial effects of pomegranate and orange juices, mainly due to their antioxidant properties, as well as the probiotic properties of kefir grains and bioactive ingredients found in kefir grains. Pomegranate: orange juice (1:1) improved the ability of kefir grains to ferment pomegranate juice and increased the survival rates of lactic acid bacteria (LAB) contained in kefir grains during storage. Specifically, 75 per cent cells remained viable after 4 weeks of storage in the fermented mixed substrate over 24 per cent in plain pomegranate juice, i.e. levels near to the limits set for products claiming probiotic properties [102].

The three commercial yeasts (for wine, beer and cider) were evaluated for the production of pomegranate alcoholic beverage (PAB) from a juice of 'Wonderful' variety at temperatures of 15 and 25°C [103]. PABs produced with these strains contained ethanol in the ranges of 5.6–7.0 per cent v/v, along with glycerol (2.65–6.05 g L^{-1}), and low volatile acidity with decrease in total flavonoid, phenolic, and monomeric anthocyanin content and free radical-scavenging activity retaining 81–91 per cent of free radical-scavenging activity, 29–41 per cent of phenolics, 24–55 per cent of flavonoids, and 66–75 per cent of anthocyanins. The 30 different volatile compounds specifically 15 esters, 4 organic acids, 8 alcohols and 3 terpenes were also reported in PABs.

A functional pomegranate beverage using Lactobacillus paracasei K5 (a potential probiotic) immobilized on delignified wheat bran was also prepared [104]. The fermentations were carried out for 24 h at pH values of 3.3, 3.6 and 3.9 followed by storage for 4 weeks at 4°C. In all pH levels, the immobilized biocatalyst was found efficient for pomegranate juice fermentation maintaining ethanol production at low levels (0.5–1 per cent v/v). It indicated the potential of L. paracasei K5 to produce good quality and synbiotic pomegranate beverages.

22. Confectionery

Pomegranate juice is processed by adding sugar and heating it to bring the thick consistency. Researchers have found that approx. 50 per cent of total anthocyanins present in the juice are lost during jelly preparation and the acidification of juice improves the colour in the final product. Anar Rub (pomegranate jam) is prepared from juice by adding sugar and heating to a thick consistency. A jelly and hard candy is prepared from pomegranate [45].

A jelly product with 55 per cent mixed fruit juice content having cultivated and wild pomegranate juices in 80:20 ratio and 45 per cent sugar was prepared [105] and found to be safely stored for a period of 6 months under both the ambient and refrigerated conditions in pet jars as well as glass jars without much change in its various chemical attributes, viz. TSS, titratable acidity, ascorbic acid, anthocyanins, pectin, total phenolics and sensory quality characteristics, viz. colour, texture, flavour and overall acceptability. The jelly formulation containing 20 per cent gelatin, pure 'Mollar de Elche' pomegranate juice, 1 per cent citric acid and sucrose as sweetener [106] produced the best quality of jellies in terms of colour (highly reddish and bright red), texture, antioxidant capacity and sensory attributes. Devhare et al. [107] formulated the 20 per cent pomegranate juice-based chikki that retained its quality attributes for 3 months in LDPE bags. Micro-encapsulated pomegranate peel extract (MEPPE) was used by Palabiyik et al. [108] in the chewing gums successfully indicating their utility as a carrier for bioactive compounds. Similarly, many scientists utilized the seed as well as the peel extracts and oils in different confectionaries. Lashkari et al. [109] prepared the whey-less fruit feta cheese containing 20 per cent pomegranate juice which had the better antioxidant properties and oxidative stability, but its texture was weaker than control.

23. Wine

The pasteurized pomegranate juice is fermented with starter wine yeast. The juice is added with sugar to adjust Brix to 22–23^0. The fermentation is allowed to continue until desired level of alcohol is obtained. Pomegranate wine contains 9–10.1 per cent of alcohol and has pH 4.0, when incubated not more than 7 days. Clarification of wine is done by bentonite treatment or by centrifugation and is aged in the same manner as red grape wine.

24. Pomegranate powder

Pomegranate arils are dried and milled to prepare low-moisture pomegranate powder. Ellagic acid is the active ingredient in pomegranate powder. The freeze-dried pomegranate powder rich in ellagic acid have many applications and may be encapsulated for regular doses [110]. Husain et al. [111] have proposed grade standards to bring the commodity under the purview of AGMARK certification. The powder is utilized in the preparation of yanggaeng, breads, jelly and other products.

25. Pomegranate seed oil

Pomegranate seed, the by-product of pomegranate juice processing, contains considerable amounts of lipid, protein, sugars and essential minerals [112, 113], higher amount of mineral elements except potassium in seed than juice ([113, 114], a range of nutraceutical components, viz. sterols, γ-tocopherol, punicic acid, hydroxyl benzoic acids [115, 116], and phenyl aliphatic glycosides as phenethyl rutinoside [116, 117]). The extracts of pomegranate seed were reported with anti-diarrhoeal and antioxidant bioactivities [116, 118]. Pomegranate seed oil is extracted by pressing pomegranate seeds. The pomegranate seeds after extracting juice contain 12–20 per cent of pomegranate seed oil. The pomegranate seed oil is rich source of conjugated octadecatrienoic fatty acids such as punicic acid. The other fatty acids include catalytic acid and alpha-eleostearic acid. Pomegranate seeds oil also had a high content of ω-3, ω-6 and ω-9 essential fatty acids in adequate quantities, which have an essential role in human nutrition, prevention of the cardiac diseases, and protection of the human body from different cancers (anticancer) and in dissolving the saturated fatty acids in the body [116]. Pomegranate seed oil can also decrease the leukotriene production from arachidonic acid that has an important role in the accuracy of asthma in children, skin inflammation and paltelet aggregation association with cardiovascular diseases [116, 119, 120]. The antioxidant capacity of pomegranate oil is less than the pomegranate peel and higher than the pomegranate juice. The pomegranate seed oil is prepared by cold-pressing the pomegranate seeds. The total crude oil content obtained from seed ranges from 13 to 15 per cent on dry matter basis. The polyunsaturated fatty acid content in the oil was reported to 87.82 per cent [121]. The encapsulation of pomegranate seed oil using lipid-based carrier system is reported to protect the sensitive compounds from degradation. It also maintains the nutritional as well as functional aspects of the pomegranate seed oil. Pomegranate seed oil commercialization is also done by nanoemulsion.

26. Pomegranate seed powder

The pomegranate peel as well as seed powder have much higher exceptionally content of lysine, isoleucine and amino acids containing sulphur (methionine and cysteine), which are usually deficient in most food stuffs, than the reference protein pattern of FAO/WHO. Seed pulp extract can be used like a natural dye in alcoholic beverages as well as to impart taste and smell [122].

Hashemi et al. [123] assessed the effects of complementary treatment with pomegranate seed powder (PSP) oral supplementation @ 5 g twice daily on patients with type 2 diabetes mellitus, and after 8 weeks of treatment, they recorded that mean differences of fasting blood glucose, glycated haemoglobin (HbA1c), total cholesterol and triglyceride (TG) were significantly decreased in the PSP group over the placebo group treatment. In addition, post-intervention values of FBG and HbA1c were significantly lower in patients treated with PSP compared to the placebo group concluding that complementary treatment with PSP may have beneficial effects on FBG and HbA1c of patients with type 2 diabetes mellitus. However, its effect on TG and cholesterol was equivocal.

27. Pomegranate waste utilization

Pomegranate plant in a whole has medicinal, industrial and cosmetic values. All parts of the pomegranate tree, i.e. roots, bark, leaves, flowers, rind and seeds, can be processed for value-added products. Rind powder is having potential uses in medicines, in leather and dye industry and is used in the preparation of tooth powder. Through advance technology developed for the preparation of rind powder, it has been found that the recovery of rind powder is 15.5 per cent on whole fruit basis and 34 per cent on rind weight basis. The pomegranate bark powder also has effective medicinal uses and is used in many *Ayurvedic* preparations.

As a part of pomegranate industrialization, the field of pomegranate by-products is an interesting field to be exploited as it is a rich source of alkaloids, aromatic compounds and enzymes.

28. Pomegranate rind powder

Extracts of pomegranate rinds provided a major source of medieval 'pomegranate' ink in Europe [124, 125]. Extracts of the rind have also been used as dyes for textiles, and specialty craft inks are still created from pomegranate [125]. Pomegranate peel comprises around 20–50 per cent of total fruit weight which is rich in minerals especially potassium, calcium, phosphorus, magnesium and sodium; complex polysaccharides and high levels of different bioactive compounds, viz. phenolics, flavonoids, proanthocyanidin compounds and ellagitannins (ETs) such as punicalagins and its isomers. It also contains the punicalin, gallagic acid, ellagic acid and ellagic acid glycosides in smaller quantities. The peel extracts being free radical inhibitors and primary antioxidants that react with free radicals suggest their potential application as natural additives and functional ingredients after its incorporation in a real food model (Heena [126]). The predominant polyphenol fractions found in pomegranate peels powder are catechins, phenol gallic acid, caffeic acid, ellagic acid, p-coumaric acid and resorcinol compounds [116]. Higher phenol content and antioxidant activity as well as being a good source of crude fibre provide numerous health benefits such as ability to improve glucose tolerance and the insulin response, decrease serum LDL-cholesterol level, and reduced hyperlipidemia and hypertension. It also contributes to gastrointestinal health and the prevention of certain cancers such as colon cancer [127]. It is also known for its apparent wound healing properties [128]. The peels are rich in antioxidants and help in keeping bacteria and other infections away. It can be used in facepacks and skincare products. Peel powder restores the pH balance of the skin and hydrates the skin by locking the moisture thus keeping it smooth, soft and supple. It also helps to protect the skin from pollutants and other environmental toxins.

A detanninated peel powder with dual benefit, i.e. separated hydrolysable tannin/ellagitannin and good amount of nutritional components with appropriate amount of tannin, can be recommended as a novel cattle feed supplement [129]. Pomegranate peel powder meal (PPPM) supplementation in a-tocopherol supplemented (POSCON) diets to broiler Cobb 500 birds promoted feed conversion ratio and protein efficiency ratio. It improved average final body and other organ weights, spleen and gizzard weights, catalase enzyme activity, and nutrient digestibility, while it reduced the concentration of serum aspartate aminotransferase at concentrations of 4–8 g/kg feed [130]. Dried pomegranate peels can be used safely in sheep feeding at 1 per cent level to realize the best growth performance and depressed the price of ration cost and best relative economic efficiency [131].

Antibacterial and antioxidant characteristics of meat burgers incorporated with pomegranate rind powder extract (PRPE) were improved due to high concentration of total phenolic content that resulted in less oxidation of lipids and exhibited potentials as appropriate natural functional substitutes for synthetic antioxidants in the high-fat meat products (Maryam [132]) as well as natural and cheap antioxidant source for the enhancement of quality of raw ground pork meat [133] in the fresh-cut food sector such as fruit salads (Valentina [134]) as well as the development of value-added food products such as bread and cookies (Tandon [135]).

29. Conclusion

Pomegranate (*P. granatum* L.) is considered as a superfruit because of its high nutritive and therapeutic values. This fruit also has high antioxidant capacity and is rich in bioactive compounds. Though pomegranate is known as food medicine and has consumer appeal, the consumption of pomegranate is still scarce as it is considered as the most difficult fruit for peeling and extraction of arils, which is time-consuming, and also the phenolic metabolites in the fruit cause irritation and hand staining during extraction. Over the period, pomegranate processing and product diversification attained thereby have played a vital role in the increased consumption and utilization of pomegranate. The research and development activities on pomegranate fruit are aimed at developing technologies for new pomegranate-derived food products. Range of products can be developed through pomegranate processing, viz. minimally processed fresh arils, juice, squash, beverage, molasses, juice concentrates, frozen seeds, jam, jelly, marmalades, grenadine, wine, seeds in syrup, pomegranate wine, pomegranate powder, pomegranate rind powder, anardana, confectionery and pomegranate seed oil. Though there is large scope in pomegranate processing, the products derived had not yet popularized on large scale only because of lack of commercially viable processing technologies. Hence, there is the need for keen and immediate attention in meeting the research and developmental gaps for the commercialization and popularization of the pomegranate products as well as pomegranate processing technology. Industrial method of peeling, standardization of extraction and proper clarification methods of pomegranate juice, development of standards for packaging and storage of pomegranate-derived products, application of the new inline technologies such as MAP, ultrafiltration for pomegranate utilization and popularization of pomegranate-based products are the aspects that required to be emphasized in the research and technology development.

Hence, experimental studies should be carried out with more informative output on the metabolic properties of pomegranate and derived products under various conditions in order to develop the scientific database and to enable the successful application of the available technology for the commercialization and utilization of pomegranate processing technology and pomegranate products.

Author details

Sangram S. Dhumal[1*], Ravindra D. Pawar[1] and Sandip S. Patil[2]

1 Horticulture Section, Rajarshee Chhatrapati Shahu Maharaj College
of Agriculture Kolhapur, Mahatma Phule Agricultural University,
Rahuri, Maharashtra, India

2 Agricultural Extension and Communication, Rajarshee Chhatrapati Shahu
Maharaj College of Agriculture Kolhapur, Mahatma Phule Agricultural University,
Rahuri, Maharashtra, India

*Address all correspondence to: sangram1326@hotmail.com

IntechOpen

References

[1] Schubert SY, Lansky EP, Neeman I. Antioxidant and eicosanoid enzyme inhibition properties of pomegranate seed oil and fermented juice flavonoids. Journal of Ethnopharmacology. 1999;**66**:11-17

[2] Saxena AK, Manan JK, Berry SK. Pomegranate post harvest technology, chemistry and processing. Indian Food Packer. 1987;**41**(7/6):43-60

[3] Anonymous. 2018. Indian Horticulture Database-2018. Pub. National Horticulture Board, Ministry of Agriculture, Govt. of India

[4] Hertog MGL, Van Popel G, Verhoeven D. Potentially anticarcinogenic secondary metabolites from fruits and vegetables. In: Tomas-Barberan FA, Robins RJ, editors. Phytochemistry of fruits and vegetables. Oxford, UK: Claredon Press; 1997. pp. 313-329

[5] Sadeghi N, Jannat B, Oveisi MR, Hajimahmoodi M, Photovat M. Antioxidant activity of Iranian pomegranate (*Punica granatum* L.) seed extracts. Journal of Agricultural Science and Technology. 2009;**11**:633-638

[6] Singh RP, Gupta AK, Bhatia AK. Utilization of wild pomegranate in north west Himalayas-Status and Problems. In: Proceedings of National Seminar on Production and Marketing of Indigenous Fruits, New Delhi. 1990. pp. 100-107

[7] Du CT, Wang PL, Francis FJ. Anthocyanins of pomegranate, *Punica granatum*. Journal of Food Science. 1975;**40**(2):417-418

[8] Gil MI, Artes F, Tomas-Barberan FA. Minimal processing and modified atmosphere packaging effects on pigmentation of pomegranate seeds. Journal of Food Science. 1996;**61**:161-164

[9] Lansky E, Hubert S, Neeman I. Pharmacological and therapeutical properties of pomegranates. In: Melgarejo P, Martinez JJ, editors. Proceedings of the First International Symposium on Pomegranate. Spain: CIHEAM; 1998. pp. 231-235

[10] Tsuda T, Watanabe M, Ohshima K, Norinobu S, Choi S, Kawakishi S. Antioxidant activity of the anthocuanin pigments cyaniding 3-O-β-D glucoside and cyaniding. Journal of Agricultural and Food Chemistry. 1994;**42**:2407-2410

[11] Malik A, Afaq F, Arfaraz S, Adhami VM, Syed DN, Mukhtar H. Pomegranate juice for chemoprevention and chemotherapy of prostate cancer. Proceedings of the National Academy of Sciences. 2005;**102**:14813-14818

[12] Sumner MD, Elliott-Eller M, Weidner G, Daubenmier JJ, Chew MH, Marlin R. Effects of pomegranate juice consumption on myocardial perfusion in patients with coronary heart disease. The American Journal of Cardiology. 2005;**96**:810-814

[13] Jurenka J. Therapeutic applications of Pomegranate: A review. Alternative Medicine Review. 2008;**13**(2):128-144

[14] Pujari KH. Studies on hardness of seeds of pomegranate (*Punica granatum* L.). A M.Sc. (Agri.) thesis submitted to Mahatma Phule Agricultural University, Rahuri, Maharashtra, India. 1983. pp. 1-85

[15] Fahan A. The seed. In: Plant anatomy. Jerusaleum: Hakkibutz hameuhad Publication; 1976. pp. 419-430

[16] Holland D, Hatib K, Bar-Ya'akov I. Pomegranate: Botany, horticulture, breeding. In: Janick J, editor. Horticultural Reveiws. Vol. 35. 2009. pp. 127-191

[17] Cristofori V, Caruso D, Latini G, Dell'Agli M, Cammilli C, Rugini E. Fruit quality of pomegranate (Punica granatum L.) autochonous varieties. European Food Research and Technology. 2011;**232**:397-403. DOI: 10.1007/s00217-010-1390-8

[18] García-Pastor ME, Zapata PJ, Castillo S, Martínez-Romero D, Guillén F, Valero D, et al. The effects of salicylic acid and its derivatives on increasing pomegranate fruit quality and bioactive compounds at harvest and during storage. Frontiers in Plant Science. 2020;**11**:668, p. 14. DOI: 10.3389/fpls.2020.00668

[19] Chavan UD, Adsule RN, Kadam SS. Physico-chemical properties of pomegranate rind powder. Beverage and Food World. 1995;**22**(1):36

[20] Ewaida EH. Nutrient composition of "Taifi" pomegranate (*Punica granatum* L.) fragments and their suitability for the production of jam. Arab Gulf Journal of Scientific Research. 1987;**3**:367-378

[21] Fadavi A, Barzegar M, Aziz H. Determination of fatty acids and total lipid content in oilseed of 25 pomegranate varieties grown in Iran. Journal of Food Composition and Analysis. 2006;**19**(6):676-680

[22] Prasad RN, Chandra R, da Silva JAT. Postharvest handling and processing of pomegranate. In: Chandra R, editor. Pomegranate, Fruit, Vegetable and Cereal Science and Biotechnology. Solapur, India: Global Science Books; 2010. pp. 88-95. (PDF) Characterization of Alternaria Species Associated with Heart Rot of Pomegranate Fruit. Available from: https://www.researchgate.net/publication/349679078_Characterization_of_Alternaria_Species_Associated_with_Heart_Rot_of_Pomegranate_Fruit [Accessed: July 6, 2022]

[23] Pantastico B. Pineapple. In: Pantastico B, editor. Postharvest, Handling and Utilization of Tropical and Subtropical Fruits and Vegetables. Westport Connecticut: AVI; 1975. pp. 65-66

[24] Pota S, Kesta S, Thongthaam MLC. Effect of packing materials and temperature on quality and storage of pomegranate fruits (Punica granatum L.). Kasestart Journal of Natural Sciences. 1987;**21**(4):328-333

[25] Prasad RN, Bankar GJ, Vashishtha BB. Effect of packaging materials on the shelf life of pomegranate fruits cv. Jalore Seedless. Progressive Horticulture. 1995;**26**:133-135

[26] Krishnamurthy S. Effect of shrink wrapping on shelf life of pomegranate. Poster Sessions Abstracts. International Food Conference-1993. 1993. pp. 150

[27] Artes F, Tomas-Barberan FA. Postharvest technological treatments of pomegranate and preparation of derived products. In: Symposium on production, processing and marketing of pomegranate in the Mediterranean region: Advances in research and technology. (Eds.) Melgarejo P, Mertinez TJ, Zaragosa, Spain: CIHEAM-IAMZ. 2000:199-204

[28] Artes F, Tudela JA, Villaescua R. Thermal postharvest treatment for improving pomegranate quality and shelf life. Postharvest Biology and Technology. 2000;**18**:245-251

[29] Waskar DP. Studies on extension of postharvest life of pomegranate fruits 'Bhagawa'. In: Sheikh MK et al., editors. Proc. IInd IS on Pomegranate and Minor, Including Mediterranean Fruits (ISPMMF - 2009). Acta Hort. 890. Vol. 2011. ISHS; 2009. pp. 455-459

[30] Ben-Arie R, Or E. The development and control of husk scald on 'Wonderful' pomegranate fruit during storage. Journal of the American Society

for Horticultural Science. 1986;**111**:395-399

[31] Artés F, Marín JG, Martínez JA. Controlled atmosphere storage of pomegranates. Zeitschrift für Lebensmittel-Untersuchung und Forschung. 1996;**203**:33-37

[32] Holcroft DM, Gil MI, Kader AA. Effect of carbon dioxide on anthocyanins, phenylalanine ammonia lyase and glucosyltransferase in the arils of stored pomegranates. Journal of the American Society for Horticultural Science. 1998;**123**:136-140

[33] Gil MI, Martinez J, Artes F. Minimally processed pomegranate seeds. Lebensmittel-Wissenschaft und Technologie. 1996;**29**:708-713

[34] Artes F. Factores de calidad y conservacion frigorifica de la Granada. In: II Jornadas Nacionales del Granado. Valencia: Univ. Politecnica de Valencia; 1992

[35] Kader AA, Chordas A, Elyatem S. Response of Pomegranates to Ethylene Treatment and Storage Temperature. USA: California Agriculture: California Agricultural Experiment Station; 1984. pp. 14-15. ISSN: 0008-0845

[36] Artes F, Marin JG, Martinez JA. Permeability rates of films for modified atmosphere packaging of respiring foods. In: Nicolai BM et al., editors. Food Quality. European Commission: Leuven; 1998. pp. 153-157

[37] Yadav K, Sarkar BC, Kumar P. Pomegranate: Recent developments in Harvesting, Processing and Utilization. Indian Food Industry. 2006;**25**(2): 55-62

[38] Lopez-Rubira V, Conesa A, Allende A, Artes F. Shelf-life and overall quality of minimally processed pomegranate arils modified atmosphere packaged and treated with UV-C.

Postharvest Biology and Technology. 2005;**37**:174-185

[39] Alper N, Savas Bahceci K, Acar J. Influence of processing and pasteurization on colour values and total phenolic compounds of pomegranate juice. Journal of Food Processing and Preservation. 2005;**29**:357-368

[40] Nuncio-Jáuregui N, Calín-Sánchez Á, Vázquez-Araujo L, Pérez-López AJ, FrutosFernández MJ, Carbonell-Barrachina ÁA. Processing pomegranates for juice and impact on bioactive components. In Processing and Impact on Active Components in Food. Chapter 76. London, UK: Elsevier; 2015. pp. 629-636

[41] Gil MI, Tomas-Barberan FA, Hess-Pierce B, Holcroft DM, Kader AA. Antioxidant activity of pomegranate juice and its relationship with phenolic composition and processing. Journal of Agricultural and Food Chemistry. 2000;**48**(10):4581-4589

[42] Singh RP, Murthy KNC, Jayaprakasha GK. Studies on the antioxidant activity of pomegranate (*Punica granatum*) peel and seed extracts using in vitro models. Journal of Agricultural and Food Chemistry. 2002;**50**(1):81-86

[43] Cemeroglu B. Nar Suyu Oretim Teknolojisi Uzerine Bir Ara-stirma. Ziraat Fak, Yayinlari, Ankara. Publ. No.664: Ankara University Press; 1977. pp. 8-28

[44] Vardin. The using possibilities of different pomegranate cultivars grown in Harran plain for using in food industry. Adana, Turkey: Ph. D. Thesis. Cukurova Univ; 2000

[45] Phadnis NA. Pomegranate for dessert and juice. Indian Horticulture. 1974;**19**(3):9

[46] Ozkan M. Degradation of anthocyanins in sour cherry and

pomegranate juices by hydrogen peroxide in the presence of added ascorbic acid. Food Chemistry. 2002;**78**:499-504

[47] de Simon BF, Perez-Ilzarbe J, Hernandez T, Gomez-Cordoves C, Estrella I. Importance of phenolic compounds for the characterization of fruit juices. Journal of Agricultural and Food Chemistry. 1992;**38**:1565-1571

[48] Spanos GA, Wrolstad RE. Phenolics of apple, pear and white grape juices and their changes with processing and storage—A review. Journal of Agricultural and Food Chemistry. 1992;**40**:1478-1487

[49] Adsule RN, Patil NB. Pomegranate: Hand Book of Fruit Science and Technology. New York: Marcel Dekker; 1995. pp. 455-464

[50] Chobe RS. Studies on extraction, clarification, preservation and storage of pomegranate (*Punica granatum* L.) juice. M.Tech. (Agri) Thesis. Rahuri (Maharashtra), India: Mahatma Phule Krishi Vidyapeeth; 1999

[51] Vardin H, Fenercioglu H. Study on the development of pomegranate juice processing technology: The pressing of pomegranate fruit. Proceedings of Ist International Symposium on Pomegranate. Ed. A. I. Ozguven. ISHS. Acta. Hort. 2009;**818**:373-381

[52] Neifar M, Ellouze-Ghorbel R, Kamoun A, Baklouti S, Mokni A, Jaouani A, et al. Effective clarification of pomegranate juice using laccase treatment optimized by response surface methodology followed by ultrafiltration. Journal of Food Process Engineering. 2011;**34**:1199-1219. DOI: 10.1111/j.1745-4530.2009.00523.x

[53] Marti N, Perez-Vicente A, Garcia-Viguera C. Influence of storage temperature and ascorbic acid addition on pomegranate juice. Journal of the Science of Food and Agriculture. 2001;**82**:217-221

[54] Kumbhar SC, Kotecha PM, Kadam SS. Effect of method of juice extraction on quality of pomegranate wine. Beverage Food World. 2002;**29**(11):35-36

[55] Sandhan VS. Preparation of carbonated beverage from pomegranate (*Punica granatum* L.) Fruits cv. Ganesh and Mridula. M.Sc. (Agri.) Thesis. Rahuri (M.S.), India: Mahatma Phule Krishi Vidyapeeth; 2003

[56] Singh J, Singh AK, Singh HK. Preparation and preservation of pomegranate (*Punica granatum* L.) RTS. Beverage and Food World. 2005;**32**(12):45-46

[57] Catania P, Alleria M, De Pasquale C, Vallone M. Effect of different processing methods on the quality of obtained pomegranate juice. In: Kalaitzis P et al., editors. Proc. III Int. Symp. on Horticulture in Europe – SHE2016. Vol. 1242. Acta Hortic; 2019. pp. 35-40. DOI: 10.17660/ActaHortic.2019. 1242.5

[58] Catania P, Comparetti A, De Pasquale C, Morello G, Vallone M. Effects of the extraction technology on pomegranate juice quality. Agronomy. 2020;**10**:1483, p. 14

[59] Sahin S, Bayindirli L, Artik N. The effect of depectinization and clarification on sour cherry juice quality. Gida. 1992;**17**(1):35-42

[60] Cemeroglu B. Meyve Suyu Uretim Teknolojisi. Ankara: Teknik Basim Sanayii; 1982

[61] Vilquez F, Laetreto C, Cooke RD. A study of the production of clarified banana juice using pectinolytic enzymes. Journal of Food Technology. 1981;**16**:115-125

[62] Vardin H, Fenercioglu H. Study on the development of pomegranate juice processing technology: Clarification of

pomegranate juice. Nahrung/Food. 2003;**47**(5):300-303

[63] Bayinderli L, Sahin S, Artik N. The effects of clarification methods on pomegranate juice quality. Fruit Processing. 1994;**9**:267-270

[64] Song-nian G, Chil X, Quing D, Ying Z. Optimization of enzymatic clarification of pomegranate juice using response surface methodology. Storage and Process. 2011;**2**:1-3

[65] Kapoor A, Iqbal H. Efficiency of tannase produced by Trichoderma Harzianum MTCC 10841 in pomegranate juice clarification and natural tannin degradation. International Journal of Biotechnology and Bioengineering Research. 2013;**4**(6):641-650

[66] Perez-Vicente A, Serrano P, Abellan P, Garcia-Viguera C. Influence of packaging material on pomegranate juice colour and bioactive compounds, during storage. Journal of the Science of Food and Agriculture. 2004;**84**:639-644

[67] Sethi V. Suitability of different packaging material for storing fruit juices and intermediate preserves. In: Souvenir of Symposium on Recent Development in Food Packaging. Mysore, India: CFTRI; 1985

[68] Waskar DP, Deshmukh AN. Effect of packaging containers on the retention of anthocyanins of pomegranate juice. Indian Food Packer. 1995;**49**(1):5-8

[69] Fathy AM. Quality and shelf life of pomegranate juice aseptically packed in different packaging materials. Middle East Journal of Applied Sciences. 2014;**4**(2):409-415

[70] Ahire DB. Studies on extraction, packaging and storage of pomegranate (*Punica granatum* L.) juice cv. Mridula. M. Sc. (Agri.). Rahuri, (Maharashtra) India: Mahatma Phule Krishi Vidyapeeth; 2007

[71] Suryawanshi AB, Kirad KS, Phad GN, Patil SB. Standardization of preservation method and their combination for safe storage of pomegranate juice at room temperature. The Asian Journal of Horticulture. 2008;**3**(2):395-399

[72] Adam JB, Ongley MH. Changes in polyphenols of red fruits during heat processing. The degradation of anthocyanins in canned fruits. Tech. Bull. 23. Chipping Campden, Gloucestershire: The Campden Food Preservation Research Association; 1972

[73] Shelar YV. Preparation of carbonated ready-to-serve (RTS) beverage from pomegranate (*Punica gramaum* L.) juice. M.Sc. (Agri.) Thesis. Rahuri (M.S.), India: Mahatma Phule Krishi Vidyapeeth; 2001

[74] Artes F, Gil MI and Martinez JA. Procedimiento para la conservacion de semillas de granada en fresco. Patent No. 9502662. 1995

[75] Sepulveda E, Galletti L, Saenz C, Tapia M. Minimal processing of pomegranate var. Wonderful. In: Melgarejo P, Mertinez TJ, editors. Symposium on Production, Processing and Marketing of Pomegranate in the Mediterranean Region: Advances in Research and Technology. Zaragosa, Spain: CIHEAM-IAMZ; 2000. pp. 237-242

[76] James Caleb O, Opara UL, Witthuhn CR. Modified atmosphere packaging of pomegranate fruit and arils: A review. Food Bioprocess Technology. 2011

[77] Ayhan Z, Esturk O. Overall quality and shelf life of minimally processed and modified atmosphere packaged "Ready-to-Eat" pomegranate arils. Journal of Food Science. 2009;**74**:C399-C405

[78] Nunes C, Graca A, Yildirim I, Sahin G, Erkan M. Metabolic response to UV-C treatment on minimally processed

pomegranate arils. In: Erkan M, Aksoy U, editors. 6th International Postharvest Symposium. Vol. 877. ISHS. Acta Horticulure; 2010. pp. 657-662

[79] Maghoumi M, Gómez P, Mostofi Y, Zamani Z, Artés-Hernández F, Artés F. Combined effect of heat treatment, UV-C and superatmospheric oxygen packing on phenolics and browning related enzymes of fresh-cut pomegranate arils. LWT—Food Science and Technology. 2014:54. DOI: 10.1016/j.lwt.2013.06.006

[80] Sepulveda E, Saenz C, Berger H, Galletti L, Valladares C, Botti C. Minimal processing of pomegranate cv. Espanola: Effect of three package material. In: Ben-Arie R, Philosoph-Hadas S, editors. Proc. 4th Int. Conf. on Postharvest. Vol. 553. ISHS; 2001. pp. 711-712

[81] Palma A, Schirra, D'Aquino S, La Malfa S, Continella G. Chemical properties changes in pomegranate seeds packaged in polypropylene trays. In: Ozguven AI, editor. Proc. Ist IS on Pomegranate. Vol. 818. ISHS; 2009. pp. 323-329

[82] Palma A, Continellab A, La Malfab S, Gentileb A, D'Aquinoa S. Overall quality of ready-to-eat pomegranate arils processed from cold stored fruit. Post Harvest Biology and Technology. 2015;**109**:1-9

[83] Hess-Pierce B, Kader A. Carbon dioxide enriched atmospheres extend postharvest life of pomegranate arils. In: Gorny JR, editor. Seventh International Controlled Atmosphere Research Conference. CA '97. Proceedings Vol.5; Fresh-Cut Fruits and Vegetables and MAP. Davis, CA: Dept. of Pomology, Univ. of California; 1997. p. 122

[84] Belay ZA, Caleb OJ, Opara UL. Impacts of low and super-atmospheric oxygen concentrations on quality attributes, phytonutrient content and

volatile compounds of minimally processed pomegranate arils (cv. Wonderful). Post Harvest Biology and Technology. 2017;**124**:119-127

[85] Belay ZA, Caleb OJ, Mahajan PV, Opara UL. Design of active modified atmosphere and humidity packaging (MAHP) for 'Wonderful' pomegranate arils. Food and Bioprocess Technology. 2018:17. DOI: 10.1007/s11947-018-2119-0. Pub. Online: May 31, 2018

[86] Ersan S, Gunes G, Zor AO. Respiration rate of pomegranate arils as affected by O_2 and CO2 and design of modified atmosphere packaging. In: Erkan M, Akoy U, editors. Proc. 10th Intern. Controlled and Modified Atmosphere Research Conference. Vol. 876. ISHS; 2010. pp. 189-196

[87] Adiletta G, Petriccione M, Liguori L, Zampella L, Mastrobuoni F, Di Matteo M. Overall quality and antioxidant enzymes of ready to eat 'Purple Queen' pomegranate arils during cold storage. Post Harvest Biology and Technology. 2019;**155**:20-28

[88] Ergun M, Ergun N. Maintaining quality of minimally processed pomegranate arils by honey treatments. British Food Journal. 2009;**111**(4):396-406

[89] Martinez-Romero D, Castillo S, Guillen F, Diaz-Mula HM, Zapata PJ, Valeroa D, et al. Aloe vera gel coating maintains quality and safety of ready-to-eatpomegranate arils. Postharvest Biology and Technology. 2013;**86**:107-112

[90] Ashtari M, Khademi O, Soufbaf M, Afsharmanesh H, Sarcheshmeh MAA. Effect of gamma irradiation on antioxidants, microbiological properties and shelf life of pomegranate arils cv. 'Malas Saveh'. Scientia Horticulturae. 2019;**244**:365-371

[91] Özdemir KS, Gökmen V. Extending the shelf-life of pomegranate arils with

chitosanascorbic acid coating. LWT-Food Science and Technology. 2017;**76**:172-180

[92] Hasheminejad N, Khodaiyan F. The effect of clove essential oil loaded chitosan nanoparticles on the shelf life and quality of pomegranate arils. Food Chemistry. 2018. DOI: 10.1016/j.foodchem.2019.125520

[93] Garcia E, Salazar DM, Melgarejo, Coret A. Determination of the respiration index and of the modified atmosphere inside the packaging of minimally processed products. In: Melgarejo P, Mertinez TJ, editors. Symposium on Production, Processing and Marketing of Pomegranate in the Mediterranean Region: Advances in Research and Technology. Zaragosa, Spain: CIHEAM-IAMZ; 2000. pp. 247-251

[94] Maestre J, Melgarejo P, Tomas-Barberan FA, Garcia-Viguera C. New food products derived from pomegranate. In: Melgarejo P, Mertinez TJ, editors. Symposium on Production, Processing and Marketing of Pomegranate in the Mediterranean Region: Advances in Research and Technology. Zaragosa, Spain: CIHEAM-IAMZ; 2000. pp. 243-245

[95] Siddappa GS, Bhatia BS. The identification of sugars in fruits by chromatography. Indian Journal of Horticulture. 1954;**11**:19

[96] Adsule RN, Kotecha PM, Kadam SS. Preparation of wine from pomegranate. Beverage and Food World. 1992;**19**(4):13-14

[97] Anonymous. Anardana. In Technology Options. FICCI Agribuisiness Information Centre. 2005. Available from: http://www.ficciagroindia.com/aic/technology-options/anardana.html

[98] Maskan M. Production of pomegranate (*Punica granatum* L.) juice concentrate by various heating methods: Colour degradation and kinetics. Journal of Food Engineering. 2006;**72**(3):218-224

[99] Hulya Orak H. Evaluation of antioxidant activity, colour and some nutritional characteristics of pomegranate (*Punica granatum* L.) juice and its sour concentrate processed by conventional evaporation. International Journal of Food Science Nutrition. 2009;**60**(1):1-11

[100] Thakur NS, Dhaygude GS, Joshi VK. Development of wild pomegranate aril-in-syrup and its quality evaluation during storage. International Journal of Food and Fermentation Technology. 2013;**3**(1):33-40. DOI: 10.5958/j.2277-9396.3.1.003

[101] Thakur M, Deshpande HW, Bhate MA. Investigation of microbial, physicochemical and color properties of probiotic pomegranate beverage during storage. International Journal of Current Microbiology and Applied Sciences. 2018;**7**:638-650

[102] Kazakos S, Mantzourani I, Nouska N, Alexopoulos A, Bezirtzoglou E, Bekatorou A, et al. Production of low-alcohol fruit beverages through fermentation of pomegranate and orange juices with kefir grains. Current Research in Nutrition and Food Science. 2016;**4**(1):19-26

[103] Kokkinomagoulos E, Nikolaou A, Kourkoutas Y, Kandylis P. Evaluation of yeast strains for pomegranate alcoholic beverage production: Effect on physicochemical characteristics, antioxidant activity, and aroma compounds. Microorganisms. 2020;**8**:1583, p. 17. DOI: 10.3390/microorganisms8101583

[104] Mantzourani I, Terpou A, Bekatorou A, Mallouchos A, Alexopoulos A, Kimbaris A, et al. Functional pomegranate beverage

production by fermentation with a novel synbiotic *L. paracasei* biocatalyst. Food Chemistry. 2019:36. DOI: 10.1016/j. foodchem.2019.125658

[105] Thakur NS, Dhaygude GS, Sharma A. Development of cultivated and wild pomegranate mixed fruit jelly and its quality evaluation during storage. Journal of Applied and Natural Science. 2017;**9**(1):587-592

[106] Cano-Lamadrid M, Calín-Sánchez A, Clemente-Villalba J, Hernández F, Carbonell-Barrachina ÁA, Sendra E, et al. Quality parameters and consumer acceptance of jelly candies based on pomegranate juice "Mollar de Elche". Foods. 2020;**9**:516. DOI: 10.3390/foods9040516

[107] Devhare PF, Kotecha PM, Godase SN, Chavan UD. Studies on utilization of pomegranate juice in the preparation of peanut chikki. International Journal of Chemical Studies. 2021;**9**(1):1532-1535

[108] Palabiyik I, Toker OS, Konar N, Gunes R, Guleri T, Alasalvar H, et al. Phenolics release kinetics in sugared and sugar-free chewing gums: Microencapsulated pomegranate peel extract usage. International Journal of Food Science and Technology. 2018:1-7

[109] Lashkari H, Varidi MJ, Eskandari MH, Varidi M. Effect of pomegranate juice on the manufacturing process and characterization of feta-type cheese during storage. Journal of Food Quality. 2020:11. Article ID 8816762. DOI: 10.1155/2020/8816762

[110] Donald EP. Ingredients. Food Technology. 2005;**59**(5):46-55

[111] Husain MI, Devikar GY, Singh G, Jaiswal PK. Quality evaluation of pomegranate seed and powder. Spice India. 2004;**17**(7):28-32

[112] El-Nemr SE, Ismail IA, Ragab M. Chemical composition of juice and seed of pomegranate fruit. Die Nahrung. 1990;**34**:601-606

[113] Dadashi S, Mousazadeh M, Emam-Djomeh Z, Mousavi SM. Pomegranate (Punica granatum L.) seed: A comparative study on biochemical composition and oil physicochemical characteristics. Biochemical composition of pomegranate seed oil. International Journal of Advanced Biological and Biomedical Research. 2013;**1**(4): 351-363

[114] El-Nemr S, Ismail I, Ragab M. The chemical composition of the juice and seeds of pomegranate fruits. Journal of Fruits Vegetables and Nuts. 1992;**11**:162-164

[115] Liu G, Xu X, Hao Q, Gao Y. Supercritical CO2 extraction optimization of pomegranate (Punica granatum L.) seed oil using response surface methodology. LWT-Food Science and Technology. 2009;**42**:1491-1495

[116] Rowayshed G, Salama A, Abul-Fadl M, Akila-Hamza S, Mohamed EA. Nutritional and chemical evaluation for pomegranate (punica granatum l.) fruit peel and seeds powders by products. Middle East Journal of Applied Sciences. 2013;**3**(4):169-179

[117] Wang RF, Xie WD, Zhang Z, Xing DM, Ding Y, Wang W, et al. Bioactive compounds from the seeds of Punica granatum (Pomegranate). Journal of Natural Products. 2004;**67**:2096-2098

[118] Singh RP, Murthy KNC, Jayaprakasha GK. Studies on the antioxidant activity of pomegranate (Punica granatum) peel and seed extracts using in vitro models. Journal of Agricultural and Food Chemistry. 2002;**50**:81-86

[119] Adhami VM, Mukhtar H. Polyphenols from green tea and pomegranate for prevention of prostate cancer. Free Radical Research. 2006;**40**:1095-1104

[120] Huxley RR, Neil H. The relationship between dietary flavonol intake and coronary heart disease mortality: A meta-analysis of prospective cohort studies. European Journal of Clinical Nutrition. 2003;**57**:904-908

[121] Kola O, Erva P, Reis AM, Devrim OB, Emrah M, Burcak T. Fatty acids, sterols and triglycerides composition of cold pressed oil from pomegranate seeds. La Rivista Italiana Delle Sostanze Grasse Vol. XCVIII. pp: 197-204. _LuGLIO/SETTEMBRE 2021. 2021. Available from: https://www.innovhub-ssi.it/kdocs/2021187/2021_vol._983_-_art._04_-_kola.pdf

[122] Pedriali CA, Fernandes AU, dos Santos PAD, da Silva MM, Severino D, da Silva MB. Ciênc. Tecnol. Aliment. Campinas. 2010;**30**(4):1017-1021

[123] Hashemi MS, Namiranian N, Tavahen H, Dehghanpour A, Rad MH, Jam-Ashkezari S, et al. Efficacy of pomegranate seed powder on glucose and lipid metabolism in patients with Type 2 Diabetes: A prospective randomized double-blind placebo-controlled clinical trial. Complementary Medicine Research. 2020;**28**:226-233. DOI: 10.1159/000510986

[124] Carvalho DN. Forty Centuries of Ink. 1999. Available from: https://www.gutenberg.org/cache/epub/1483/pg1483-images.html [Accessed: October 6, 2021]

[125] Stover E, Mercure EW. The pomegranate: A new look at the fruit of paradise. HortScience. 2007;**42**(5): 1088-1092

[126] Jalal H, Pal MA, Hamdani H, Rovida M, Khan NN. Antioxidant activity of pomegranate peel and seed powder extracts. Journal of Pharmacognosy and Phytochemistry. 2018;**7**(5):992-997

[127] Ranjitha J, Bhuvaneshwari G, Terdal D, Kavya K. Nutritional composition of fresh pomegranate peel powder. International Journal of Chemical Studies. 2018;**6**(4):692-696

[128] Murthy KN, Reddy VK, Veigas JM, Murthy UD. Study on wound healing activity of Punica granatum peel. Journal of Medicinal Food. 2004;**7**(2):256-259. DOI: 10.1089/1096620041224111

[129] Kushwaha SC, Bera MB, Kumar P. Nutritional composition of detanninated and fresh pomegranate peel powder. IOSR Journal Of Environmental Science, Toxicology And Food Technology. 2013;**7**(1):38-42

[130] Akuru EA, Mpendulo CT, Oyeagu CE, Nantapo CWT. Pomegranate (Punica granatum L.) peel powder meal supplementation in broilers: Effect on growth performance, digestibility, carcase and organ weights, serum and some meat antioxidant enzyme biomarkers. Italian Journal of Animal Science. 2021;**20**(1):119-131. DOI: 10.1080/1828051X.2020.1870877

[131] Omer HAA, Abdel-Magid SS, Awadalla IM. Nutritional and chemical evaluation of dried pomegranate (Punica granatum L.) peels and studying the impact of level of inclusion in ration formulation on productive performance of growing Ossimi lambs. Bulletin of the National Research Centre. 2019;**43**:182. DOI: 10.1186/s42269-019-0245-0

[132] Shahamirian M, Eskandari MH, Niakousari M, Esteghlal S, Gahruie HH, Khaneghah KM. Incorporation of pomegranate rind powder extract and pomegranate juice into frozen burgers: Oxidative stability, sensorial and microbiological characteristics. Journal of Food Science and Technology.

2019;**56**(3):1174-1183. DOI: 10.1007/
s13197-019-03580-5

[133] Qin Y-Y, Zhang Z-H, Li L, Xiong W,
Shi J-Y, Zhao T-R, et al. Antioxidant
effect of pomegranate rind powder
extract, pomegranate juice and
pomegranate seed powder extract as
antioxidants in raw ground pork meat.
Food Science and Biotechnology.
2013;**22**(4):1063-1069. DOI: 10.1007/
s10068-013-0184-8

[134] Lacivita V, Incoronato AL,
Conte A, Del Nobile MA. Pomegranate
peel powder as a food preservative in
fruit salad: A sustainable approach.
Food. 2021;**2021**(10):1359. DOI:
10.3390/foods10061359

[135] Palak T, Ranu P, Anisha V. Sensory
Analysis of Pomegranate Peel Powder in
the Development of Value Added Food
Products. International Journal of
Science and Research. 2020;**9**(11):414-
417. DOI: 10.21275/SR201106190108